AVAILABILITY OF REFERENCE MATERIALS
IN NRC PUBLICATIONS

NUREG-1556
Vol. 14

Consolidated Guidance About Materials Licenses

Program-Specific Guidance About Well Logging, Tracer, and Field Flood Study Licenses

Final Report

Manuscript Completed: June 2000
Date Published: June 2000

Prepared by
J.E. Whitten, S.R. Courtemanche, A.R. Jones,
R.E. Penrod, D.B. Fogle

Division of Industrial and Medical Nuclear Safety
Office of Nuclear Material Safety and Safeguards
U.S. Nuclear Regulatory Commission
Washington, DC 20555-0001

ABSTRACT

As part of its redesign of the materials licensing process, NRC is consolidating and updating numerous guidance documents into a single comprehensive repository as described in NUREG-1539, "Methodology and Findings of the NRC's Materials Licensing Process Redesign," dated April 1996, and draft NUREG-1541, "Process and Design for Consolidating and Updating Materials Licensing Guidance," dated April 1996. NUREG-1556, Vol. 14, "Consolidated Guidance about Materials Licenses: Program-Specific Guidance about Well Logging, Tracer, and Field Flood Study Licenses," dated June 2000, is the fourteenth program-specific guidance document developed for the new process and is intended for use by applicants, licensees, and NRC staff, and will also be available to Agreement States. This document combines and updates the guidance found in Draft Regulatory Guide, "Guide for the Preparation of Applications for the Use of Radioactive Materials in Well Logging Operations," dated July 1987. This report takes a more risk-informed, performance-based approach to licensing of well logging, tracer, and field flood study operations, and it reduces the information (amount and level of detail) needed to support an application for these activities.

CONTENTS

CONTENTS

APPENDICES

CONTENTS

FIGURES

TABLES

CONTENTS

FOREWORD

The United States Nuclear Regulatory Commission (NRC) is using Business Process Redesign (BPR) techniques to redesign its materials licensing process. This effort is described in NUREG-1539, "Methodology and Findings of the NRC's Materials Licensing Process Redesign," dated April 1996. A critical element of the new process is consolidating and updating numerous guidance documents into a NUREG series of reports. Below is a list of volumes currently included in the NUREG-1556 series:

Vol. No.	Volume Title	Status
1	Program-Specific Guidance About Portable Gauge Licenses	Final Report
2	Program-Specific Guidance About Industrial Radiography Licenses	Final Report
3	Applications for Sealed Source and Device Evaluation and Registration	Final Report
4	Program-Specific Guidance About Fixed Gauge Licenses	Final Report
5	Program-Specific Guidance About Self-Shielded Irradiator Licenses	Final Report
6	Program-Specific Guidance About 10 CFR Part 36 Irradiator Licenses	Final Report
7	Program-Specific Guidance About Academic, Research and Development, and Other Licenses of Limited Scope	Final Report
8	Program-Specific Guidance About Exempt Distribution Licenses	Final Report
9	Program-Specific Guidance About Medical Use Licenses	Draft for Comment
10	Program-Specific Guidance About Master Materials Licenses	Draft for Comment
11	Program-Specific Guidance About Licenses of Broad Scope	Final Report
13	Program-Specific Guidance About Commercial Radiopharmacy Licenses	Final Report
14	Program-Specific Guidance About Well Logging, Tracer, and Field Flood Study Licenses	Final Report
15	Guidance About Changes of Control and About Bankruptcy Involving Byproduct, Source, or Special Nuclear Materials Licenses	Draft for Comment
16	Program-Specific Guidance About Licenses Authorizing Distribution to General Licensees	Draft for Comment

Vol. No.	Volume Title	Status
17	Program-Specific Guidance About Licenses for Special Nuclear Material of Less Than Critical Mass	Draft for Comment
18	Program-Specific Guidance About Service Provider Licenses	Draft for Comment
19	Guidance for Agreement State Licensees About NRC Form 241, "Report of Proposed Activities in Non-Agreement States, Areas of Exclusive Federal Jurisdiction, or Offshore Waters" and Guidance For NRC Licensees Proposing to Work in Agreement State Jurisdiction (Reciprocity)	Draft for Comment
20	Guidance About Administrative Licensing Procedures	Draft for Comment

The current document, NUREG-1556, Vol. 14, "Consolidated Guidance about Materials Licenses: Program-Specific Guidance about Well Logging, Tracer, and Field Flood Study Licenses," dated June 2000, is the fourteenth program-specific guidance document developed for the new process. It is intended for use by applicants, licensees, NRC license reviewers, and other NRC personnel. It combines and updates the guidance for applicants and licensees previously found in a working draft of a "Guide for the Preparation of Applications for the Use of Radioactive Materials in Well Logging Operations," dated July 1987. In addition, this report also contains pertinent information found in NUREG Reports, Regulations, Guides, Policy and Guidance Directories, Information Notices, and other documents as listed in Appendix A.

This report takes a risk-informed, performance-based approach to licensing well logging, tracer, and field flood study applications. It reduces the amount of information needed from an applicant seeking to possess and use radioactive materials in these applications.

A team composed of NRC staff from Headquarters, the Regional Offices, and Agreement State representatives from Louisiana and Texas drafted this document, drawing on their collective experience in radiation safety in general and as specifically applied to well logging, tracer, and field flood study operations. A representative of NRC's Office of the General Counsel provided a legal perspective.

This document presents a step in the transition from the current paper-based process to the new electronic process. It is available on the Internet at the following address: <http://www.nrc.gov/NRC/NUREGS/SR1556/V14/index.html>.

This document is not a substitute for NRC regulations, and compliance is not required. The approaches and methods described in this report are provided for information only. Methods and solutions different from those described in this report will be acceptable if they provide a basis for the staff to make the determination needed to issue or continue a license.

Donald A. Cool, Director
Division of Industrial and Medical Nuclear Safety
Office of Nuclear Material Safety and Safeguards

ACKNOWLEDGMENTS

The writing team thanks the individuals listed below for assisting in the development and review of this report. All participants provided valuable insights, observations, and recommendations.

The team thanks Dianne Geshen, Rolonda Jackson, Tamra King, D.W. Benedict Llewellyn, and Agi Seaton, of Computer Sciences Corporation.

Additionally, the team would like to acknowledge and thank Dwaine Brown of Halliburton Energy Services, Ken Turner of Schlumberger Technology, and Phil Stoehr of Western Atlas International, for providing photographs, graphics, text review, and technical input.

The Participants

Brown, Carrie
Cain, Charles L.
Caniano, Roy J.
Cardwell, Cynthia C.
Collins, Douglas M.
Courtemanche, Steven R.
Camper, Larry W.
Combs, Frederick, C
Fogle, David B.
Haisfield, Mark F.
Jones, Andrea R.
Joustra, Judith A.
Leonardi, Richard A.
Merchant, Sally L.
Penrod, Richard E.
Schwartz, Maria E.
Spitzberg, D. Blair
Treby, Stuart A.
Whitten, Jack E.

ABBREVIATIONS

ALARA	as low as is reasonably achievable
ALI	Annual Limit on Intakes
ANSI	American National Standards Institute
bkg	background
BPR	business process redesign
Bq	becquerel
CDE	committed dose equivalent
CEDE	committed effective dose equivalent
CFR	Code of Federal Regulations
C/kg	coulombs/kilogram
cpm	counts per minute
DFP	Decommissioning Funding Plan
DIS	decay-in-storage
DOE	United States Department of Energy
DOT	United States Department of Transportation
dpm	disintegrations per minute
DTS	drill-to-stop
EA	environmental assessment
ECS	energy compensation source
EDE	effective dose equivalent
EPA	United States Environmental Protection Agency
F/A	financial assurance
FDA	United States Food and Drug Administration
FR	Federal Register
G-M	Geiger-Mueller
GBq	gigabecquerel
GPO	Government Printing Office
IN	Information Notice
LLW	low level waste
LSA	low specific activity
LWD	logging while drilling
MBq	megabecquerel
MC	Manual Chapter
mGy	milligray
mR	milliroentgen
mrem	millirem

ABBREVIATIONS

mSv	millisievert
MWD	measurement while drilling
NCRP	National Council on Radiation Protection and Measurements
NIST	National Institute of Standards and Technology
NMSS	Office of Nuclear Material Safety and Safeguards
NORM	naturally-occurring radioactive material
NRC	United States Nuclear Regulatory Commission
NVLAP	National Voluntary Laboratory Accreditation Program
OCFO	Office of the Chief Financial Officer
OCR	optical character reader
OGC	Office of the General Counsel
OMB	Office of Management and Budget
OSP	Office of State Programs
OSL	optically stimulated luminescence
QA	quality assurance
R	roentgen
RES	Office of Nuclear Regulatory Research
RG	Regulatory Guide
RQ	reportable quantities
RSO	radiation safety officer
SDE	shallow dose equivalent
SI	International System of Units (abbreviated SI from the French Le Systeme Internationale d'Unites)
SSD	sealed source and device
std	standard
Sv	sievert
T1/2	Half-life
TAR	technical assistance request
TEDE	total effective dose equivalent
TI	transportation index
TLD	thermoluminescent dosimeters
USASI	United States of America Standards Institute
USC	United States Code
USDA	United States Department of Agriculture

1 PURPOSE OF REPORT

Byproduct material, as defined in 10 CFR 30.4, depleted uranium, as defined in 10 CFR 40.4, and special nuclear material, as defined in 10 CFR 70.4, are used for a variety of purposes to include: well logging and tracer applications involving both single or multiple well bores; conventional well logging and tracer operations; and, in some cases, research and development. Examples include the following applications:

- Sealed sources are used in cased and uncased boreholes

- Tracer materials are used in single well applications

- Tracer materials are used in multiple well applications (field flood study) for enhanced recovery of oil and gas wells

- Sealed sources are used for calibration of applicant's survey instruments and well logging tools

- Sealed sources and tracer materials are used in the research and development of new techniques and equipment.

This report provides guidance to an applicant in preparing a well logging, tracer, and field flood study license application as well as NRC criteria for evaluating the corresponding license application. It identifies the information needed to complete NRC Form 313 (Appendix B), Application for Material License, for the use of sealed byproduct materials in well logging, and unsealed byproduct materials in tracer and field flood study applications. The information collection requirements in 10 CFR Parts 19, 20, 21,30, 32, 39, 40, 51, 70, and NRC Form 313 have been approved under the Office of Management and Budget (OMB) Control Numbers 3150-0044, 3150-014, 3150-35, 3150-0130, 3150-0017, 3150-0001, 3150-0130, 3150-0020, 3150-0021, 3150-0009, and 3150-0120.

The format within this document for each item of technical information is as follows:

- Regulations - references the regulations applicable to the item

- Criteria - outlines the criteria used to judge the adequacy of the applicant's response

- Discussion - provides additional information on the topic sufficient to meet the needs of most readers, and

- Response from Applicant - provides suggested response(s), offers the option of an alternative reply, or indicates that no response is needed on that topic during the licensing process.

Notes and References are self-explanatory.

NRC Form 313 does not have sufficient space for applicants to provide full responses to Items 5 through 11; as indicated on the form, the answers to those items are to be provided on separate sheets of paper and submitted with the completed NRC Form 313. For the convenience of applicants and for streamlined handling of applications for well logging, tracer, or field flood study licenses, use Appendix C to provide supporting information, attach it to NRC Form 313, and submit it to NRC.

Appendix D is a checklist that NRC staff use to review applications and that applicants can use to check for completeness. Appendix E is a sample well logging license, containing the conditions most often found on these licenses, although not all licenses will have all conditions. Appendices F through V contain additional information on various radiation safety topics.

Appendix F provides specific guidance for licensing field flood activities.

In this document, "dose" or "radiation absorbed dose" includes: dose equivalent; effective dose equivalent (EDE); committed dose equivalent (CDE); committed effective dose equivalent (CEDE); or total effective dose equivalent (TEDE). These terms are defined in 10 CFR Part 20. Rem, and its SI [Systeme International-(international units)] equivalent Sievert [1 rem = 0.01 Sievert (Sv)], is used to describe units of radiation exposure or dose.

2 AGREEMENT STATES

Certain states, called Agreement States (see Figure 2.1), have entered into agreements with the NRC that give them the authority to license and inspect byproduct, source, or special nuclear materials used or possessed within their borders. Any applicant, other than a Federal Agency, who wishes to possess or use licensed material in one of these Agreement States needs to contact the responsible officials in that State for guidance on preparing an application; file these applications with State officials, not with the NRC.

In general, NRC's materials licensees who wish to conduct operations under reciprocity at temporary jobsites in an Agreement State should contact that State's radiation control program office for information about State regulations. To ensure compliance with Agreement State reciprocity requirements, a licensee should request authorization well in advance of scheduled use.

Under the provisions of 10 CFR 150.20, NRC can recognize and grant a general license to Agreement State licensees. This general license authorization allows Agreement State licensees to conduct licensed operations identified on the Agreement State license in Non-Agreement States; areas of exclusive Federal jurisdiction within Agreement States; and offshore waters provided:

- The Agreement State license does not limit authorized activity to a specific installation or location

- The Agreement State license contains no provisions to the contrary

- Activities, other than those in offshore waters, including storage of materials, are limited to a total of 180 days in any calendar year. Offshore activities, as specified in 10 CFR 150.20(b)(4), are authorized for an unlimited period of time

- NRC must be notified in accordance with the provision of 10 CFR 150.20(b)(1).

Licensees who are requesting generally licensed activities in offshore waters off of Louisiana, and are licensed by the State of Louisiana, can notify the State of Louisiana in lieu of notifying NRC. Notification to the State of Louisiana must be completed in accordance with the provisions of 10 CFR 150.20(c).

In the special situation of work at Federally-controlled sites in Agreement States, it is necessary to know the jurisdictional status of the land in order to determine whether NRC or the Agreement State has regulatory authority. As indicated above, NRC has regulatory authority only over land determined to be "exclusive Federal jurisdiction," while the Agreement State has jurisdiction over non-exclusive Federal jurisdiction land. Licensees are responsible for finding out, in advance, the jurisdictional status of the specific areas where they plan to conduct licensed operations. NRC recommends that licensees ask their local contact for the Federal Agency controlling the site (e.g., contract officer, base environmental health officer, district office staff) to help determine the jurisdictional status of the land and to provide the information in writing, so that licensees can comply with NRC or Agreement State regulatory requirements, as appropriate. Additional guidance on determining jurisdictional status is found in All Agreement States Letter, SP-96-022, dated February 16, 1996, which is available from NRC upon request.

Table 2.1 provides a quick way to check on which Agency has regulatory authority.

Table 2.1   Who Regulates the Activity?

Applicant and Proposed Location of Work	Regulatory Agency
Federal Agency, regardless of location (except that Department of Energy [DOE] and, under most circumstances, its prime contractors are exempt from licensing [10 CFR 30.12])	NRC
Non-Federal entity in non-Agreement State, US territory, or possession	NRC
Non-Federal entity in Agreement State at non-Federally controlled site	Agreement State
Non-Federal entity in Agreement State at Federally-controlled site *not* subject to exclusive Federal jurisdiction	Agreement State
Non-Federal entity in Agreement State at Federally-controlled site subject to exclusive Federal jurisdiction	NRC

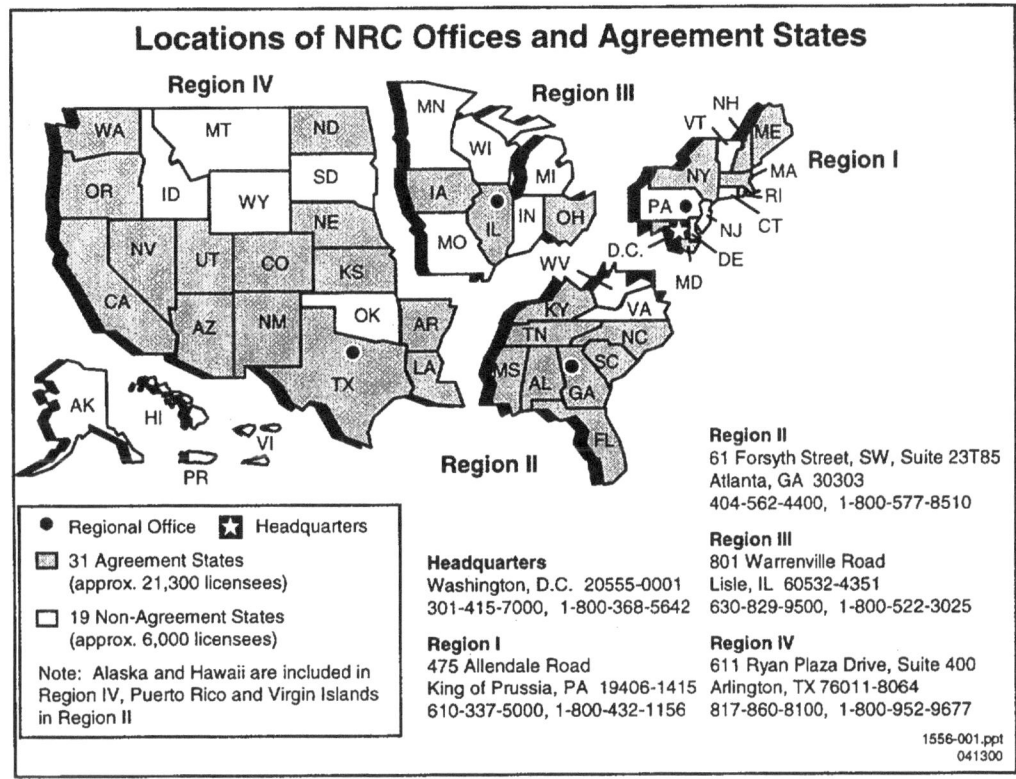

Figure 2.1 U.S. Map. *Location of NRC Offices and Agreement States.*

Reference: A current list of Agreement States (including names, addresses, and telephone numbers of responsible officials) may be obtained upon request from NRC's Regional Offices. Or visit NRC's Home Page <http://www.nrc.gov>, choose "Nuclear Materials," then "Review of State Radiation Control Program Query Form," and then "Directories."

The All Agreement States Letter, SP-96-022, dated February 16, 1996, is available by contacting NRC's Office of State Programs; call NRC's toll free number (800) 368-5642, and then ask for extension 415-3340. Or visit NRC's Home Page <http://www.nrc.gov>, choose "Nuclear Materials," then choose "Review of State Radiation Control Program Query Form" and follow the directions for submitting a query for "SP-96-022."

3   MANAGEMENT RESPONSIBILITY

The NRC recognizes that effective radiation safety program management is vital to achieving safe and compliant operations. NRC believes that consistent compliance with its regulations provides reasonable assurance that licensed activities will be conducted safely. NRC also believes that effective management will result in increased safety and compliance.

> Management refers to the processes for conducting and controlling radiation safety programs and to the individuals who are responsible for those processes and who have the authority to provide necessary resources to achieve regulatory compliance.

To ensure adequate management involvement, a management representative must sign the submitted application acknowledging management's commitments and responsibility for the following:

- Radiation safety, security and control of radioactive materials, and compliance with regulations

- Completeness and accuracy of the radiation safety records and all information provided to NRC (10 CFR 30.9)

- Knowledge about the contents of the license and application

- Compliance with current NRC and Department of Transportation (DOT) regulations and the licensee's operating and emergency procedures

- Commitment to provide adequate resources (including space, equipment, personnel, time, and, if needed, contractors) to the radiation protection program to ensure that the public and workers are protected from radiation hazards and that compliance with the regulations is maintained

- Selection and assignment of a qualified individual to serve as the Radiation Safety Officer (RSO) for licensed activities

- Prohibition against discrimination of employees engaged in protected activities (10 CFR 30.7)

- Commitment to provide information to employees regarding the employee protection, completeness and accuracy of information, and deliberate misconduct provisions in 10 CFR 30.7, 10 CFR 30.9 and 10 CFR 30.10

- Obtaining NRC's prior written consent before transferring control of the license

- Notifying the appropriate NRC Regional Administrator in writing, immediately following filing of petition for voluntary or involuntary bankruptcy.

MANAGEMENT RESPONSIBILITY

For information on NRC inspection, investigation, enforcement, and other compliance programs, see the current version of "General Statement of Policy and Procedures for NRC Enforcement Actions," NUREG-1600, and Manual Chapter (MC) 87113, Appendix G, "Suggested Well Logging, Tracer, and Field Flood Study Audit Checklist." NUREG-1600 is available electronically at <http://www.nrc.gov/OE>. For hard copies of NUREG-1600 and MC 87113, see the Notice of Availability (on the inside front cover of this report).

4 APPLICABLE REGULATIONS

It is the applicant's or licensee's responsibility to have up-to-date copies of applicable regulations, read them, understand them, and comply with each applicable regulation.

The following Parts of 10 CFR Chapter I contain regulations applicable to well logging, tracer, and field flood study licenses:

- 10 CFR Part 2, "Rules of Practice for Domestic Licensing Proceedings and Issuance of Orders"

- 10 CFR Part 19, "Notices, Instructions and Reports to Workers: Inspection and Investigations"

- 10 CFR Part 20, "Standards for Protection Against Radiation"

- 10 CFR Part 21, "Reporting of Defects and Noncompliance"

- 10 CFR Part 30, "Rules of General Applicability to Domestic Licensing of Byproduct Material"

- 10 CFR Part 32, "Specific Domestic Licenses to Manufacture or Transfer Certain Items Containing Byproduct Material"

- 10 CFR Part 33, "Specific Domestic Licenses of Broad Scope for Byproduct Material"

- 10 CFR Part 39, "Licenses and Radiation Safety Requirements for Well Logging"

- 10 CFR Part 40, "Domestic Licensing of Source Material"

- 10 CFR Part 70, "Domestic Licensing of Special Nuclear Material"

- 10 CFR Part 71, "Packaging and Transportation of Radioactive Material."

> Part 71 requires that licensees or applicants who transport licensed material outside the site of usage, as specified in the NRC license, or where transport is on public highways, or who deliver licensed material to a carrier for transport, shall comply with the applicable requirements of the DOT that are found in 49 CFR Parts 170 through 189, appropriate to the mode of transport. Copies of DOT regulations can be ordered from the Government Printing Office (GPO), whose address and telephone number are listed below.

- 10 CFR Part 110, "Export and Import of Nuclear Equipment and Material"

- 10 CFR Part 150, "Exemptions and Continued Regulatory Authority in Agreement States and in Offshore Waters Under Section 274"

- 10 CFR Part 170, "Fees for Facilities, Materials, Import and Export Licenses and Other Regulatory Services Under the Atomic Energy Act of 1954, as Amended"

- 10 CFR Part 171, "Annual Fees for Reactor Operating Licenses, and Fuel Cycle Licenses and Materials Licenses, Including Holders of Certificates of Compliance, Registrations, and Quality Assurance Program Approvals and Government Agencies Licensed by NRC."

To request copies of the above documents, call GPO's order desk in Washington, DC at (202) 512-1800. Order the two-volume bound version of *Title 10, Code of Federal Regulations,*

Parts 0-50 and 51-199 from the GPO, Superintendent of Documents, Post Office Box 371954, Pittsburgh, Pennsylvania 15250-7954. You may also contact the GPO electronically at <http://www.gpo.gov>. Additionally, Title 10, Code of Federal Regulations, Parts 0-50 and 51-199, is available electronically from NRC's reference library at <http://www.nrc.gov>. Individuals may request single hard copies of the above documents from NRC's Regional Offices (see Figure 2.1 for addresses and telephone numbers). Note that NRC publishes amendments to its regulations in the *Federal Register*.

5 HOW TO FILE

5.1 PAPER APPLICATION

Applicants for a materials license should do the following:

- Be sure to use the most recent guidance in preparing an application

- Complete NRC Form 313 (Appendix B) Items 1 through 4, 12, and 13 on the form itself

- Complete NRC Form 313 Items 5 through 11 on supplementary pages, or use Appendix C

- For each separate sheet that is submitted with the application, other than Appendix B, identify and key it to the item number on the application or the topic to which it refers

- Submit all documents, including drawings, if practicable, on 8-1/2 x 11 inch paper. If the submission of larger documents is necessary, subdivide the document into 8-1/2 x 11 inch pages so that it can be reassembled by the NRC staff when required.

- Identify each drawing with drawing number, revision number, title, date, scale, and applicant's name. Clearly indicate if drawings have been reduced or enlarged.

- Avoid submitting proprietary information unless it is necessary

- Submit an original, signed application and one copy

- Retain one copy of the license application for future reference.

> **As required by 10 CFR 30.32(c), applications shall be signed by a duly authorized management representative; see section on "Certification."**

> **Using the suggested wording of responses and committing to using the model procedures in NUREG-1556, Vol. 14 will expedite NRC's review.**

All license applications will be available for review by the general public in NRC's Public Document Rooms. If it is necessary to submit proprietary information, follow the procedure in 10 CFR 2.790. Failure to follow this procedure could result in disclosure of the proprietary information to the public or substantial delays in processing the application. Employee personal information, i.e., home address, home telephone number, Social Security Number, date of birth, and radiation dose information, should not be submitted unless specifically requested by NRC.

- Do not submit personal information about employees

- Do not submit copies of NRC licenses.

As explained in the "Foreword," NRC's new licensing process will be faster and more efficient, in part, through acceptance and processing of electronic applications at some future date. NRC will continue to accept paper applications; however, these will be scanned and converted to an electronic format. To ensure a smooth transition, applicants are requested to follow these suggestions:

- Choose typeface designs that are sans serif, such as Arial, Helvetica, Futura, and Universe; the text of this document is in a serif font called Times New Roman

- Submit printed or typewritten, not handwritten, text on smooth, crisp paper that will feed easily into the scanner

- Choose 12-point or larger font size

- Avoid stylized characters such as script, italic, etc.

- Be sure the print is clear and sharp

- Be sure there is high contrast between the ink and paper (black ink on white paper is best).

5.2 ELECTRONIC APPLICATION

As the electronic licensing process develops, it is anticipated that NRC may provide mechanisms for filing applications via diskettes, CD-ROM, and the Internet. Additional filing instructions will be provided as these new mechanisms become available. The existing paper process will be used until such time.

6 WHERE TO FILE

Applicants wishing to possess or use licensed material in any State or U. S. Territory or possession subject to NRC jurisdiction must file an application with the NRC Regional Office for the locale in which the material will be possessed and/or used. Figure 2.1 shows NRC's four Regional Offices and their respective areas for licensing purposes and identifies Agreement States.

In general, applicants wishing to possess or use licensed material in Agreement States must file an application with the Agreement State, not NRC. However, if work will be conducted at Federally-controlled sites in Agreement States, applicants must first determine the jurisdictional status of the land in order to determine whether NRC or the Agreement State has regulatory authority. See the section on "Agreement States" for additional information.

7   LICENSE FEES

Each application for which a fee is specified must be accompanied by the appropriate fee. Refer to 10 CFR 170.31 to determine the amount of the fee. NRC will not issue the new license prior to fee receipt. An application for a new license or an amendment to an existing license requesting authorization to conduct field flood studies requires that an environmental assessment be performed. Fees for a licensing action that requires an environmental assessment are charged at an hourly rate. Full cost fee recovery is assessed by the professional staff time expended, as described in footnote e.3. to 10 CFR 170.31. Once technical review begins, no fees will be refunded; application fees will be charged regardless of NRC's disposition of an application or the withdrawal of an application.

Most NRC licensees are also subject to annual fees; refer to 10 CFR 171.16. Consult 10 CFR 171.11 for additional information on exemptions from annual fees and 10 CFR 171.16(c) on reduced annual fees for licensees that qualify as "small entities."

Direct all questions about NRC's fees or completion of Item 12 of NRC Form 313 (Appendix B) to the Office of the Chief Financial Officer (OCFO) at NRC Headquarters in Rockville, Maryland, (301) 415-7554. You may also call NRC's toll free number, (800) 368-5642, and then ask for extension 415-7554.

8 CONTENTS OF AN APPLICATION

The following comments apply to the indicated items on NRC Form 313 (Appendix B).

8.1 ITEM 1: LICENSE ACTION TYPE

THIS IS AN APPLICATION FOR (Check appropriate item):

Type of Action	License No.
[ ] A. New License	Not Applicable
[ ] B. Amendment	XX-XXXXX-XX
[ ] C. Renewal	XX-XXXXX-XX

Check box A for a new license request.

Check box B for an amendment to an existing license, and provide license number.

Check box C for a renewal of an existing license, and provide license number.

8.2 ITEM 2: APPLICANT'S NAME AND MAILING ADDRESS

Response from Applicant: List the legal name of the applicant's corporation or other legal entity with direct control over use of the radioactive material; a division or department within a legal entity may not be a licensee. An individual may be designated as the applicant only if the individual is acting in a private capacity and the use of the radioactive material is not connected with employment in a corporation or other legal entity. Provide the mailing address where correspondence should be sent. A Post Office box or drawer number is an acceptable mailing address.

Notify NRC of changes in mailing address; these changes do not require a fee.

Note: NRC must be notified before control of the license is transferred or when bankruptcy proceedings have been initiated. See below for more details. NRC Information Notice (IN) 97-30, "Control of Licensed Material during Reorganizations, Employee-Management Disagreements, and Financial Crises," dated June 3, 1997, discusses the potential for the security and control of licensed material to be compromised during periods of organizational instability.

.

Granting of an NRC license does not relieve a licensee from complying with other applicable Federal, State, or local regulations (e.g., local zoning requirements or a local ordinance requiring registration of a radiation-producing device).

8.4 ITEM 4: PERSON TO BE CONTACTED ABOUT THIS APPLICATION

Identify the name and title of the individual who can answer questions about the application and include his or her telephone number. This is typically the proposed RSO or a principal user of radioactive materials, unless the applicant has named a different person as the contact. The NRC will contact this individual if there are questions about the application.

Notify NRC if the contact person or the contact person's telephone number changes so that NRC can contact the applicant or licensee in the future with questions, concerns, or information. This notice is for "information only" and does not require a license amendment or a fee.

As indicated on NRC Form 313 (Appendix B), Items 5 through 11 should be submitted on separate sheets of paper. Applicants may use Appendix C for this purpose and should note that using the suggested wording of responses and committing to using the model procedures in this report will expedite NRC's review.

8.5 ITEM 5: RADIOACTIVE MATERIAL

Regulations: 10 CFR 30.18, 10 CFR 30.32(g), 10 CFR 30.32(i), 10 CFR 30.33(a)(2), 10 CFR 32.210, 10 CFR 39.13.

Criteria: An application for a license will be approved if the requirements of 10 CFR 30.33 and 10 CFR 39.13 are met. In addition, licensees will be authorized to possess and use only those sealed sources and devices that are specifically approved or registered by NRC or an Agreement State.

Discussion: The applicant should list each requested radioisotope by its element name and mass number (e.g., cesium-137), specify whether the material will be acquired and used in unsealed or sealed form, and list the maximum amount requested. Table 8.1 below provides examples of the different types of radioactive materials. Some, not all, are addressed in this report.

Note: Additional safety equipment and precautions are required when handling and using unsealed free-form volatile radioactive materials. (Volatile means that a liquid becomes a gas at a relatively low temperature when the sealed container within which the liquid is stored is left open to the environment.)

Table 8.1   Types of Radioactive Materials

Type of Material	Covered by this Report	Examples
Byproduct (reactor-produced)	Yes	H-3, C-14, Na-22, S-35, Sc-46, Ca-45, *Fe-55, *Co-57, Co-60, Ni-63, Zn-65, Br-82, Sr-85, Sr-90, Ag-110m, I-125, Sb-124, I-131, Xe-133, Cs-137, La-140, Ir-192, Au-198, Am-241
Source material (Depleted Uranium)	Yes	Depleted Uranium
Special nuclear material	Yes	Pu-238:Be Sealed Source
Naturally occurring radioisotopes	No	Ra-226, Th-232, Th Natural
Accelerator-produced radioisotopes	No	Na-22, *Fe-55, *Co-57, Co-58

* Both accelerator and reactor produced

Possession limits should be specified in megabecquerels (MBq) [millicuries (mCi)] or gigabecquerels (GBq) [curies (Ci)] for each radioisotope. Applicants should include in the possession limits requested the total estimated inventory, including licensed material in storage and maintained as radioactive waste. The requested possession limits for any radioisotope should be commensurate with the applicant's needs and facilities for safe handling. Applicants, when establishing their possession limits for radioactive materials with half lives greater than 120 days, should review the requirements for submitting a certification for financial assurance for decommissioning. These requirements are discussed in the Section on Financial Assurance and Decommissioning and in Appendix I.

Applicants requesting an authorization to use volatile radioactive material must provide appropriate facilities, handling equipment, and radiation safety procedures for using such material.

If a dose evaluation indicates, due to a release of radioactive materials, that the potential dose to a person off-site would exceed 0.01 sieverts (Sv)[1 rem] effective dose equivalent or 0.05 Sv [5 rems] to the thyroid, an emergency plan for responding to a release shall be included with the application. For iodine-131, the quantity requiring an emergency plan is 370 GBq [10 curies].

For non-federal licensees, requests to license naturally-occurring radioactive material (NORM) and accelerator-produced radioactive material should be made to the appropriate State regulatory Agency. NRC does not regulate NORM or accelerator-produced radioactive material.

Consult with the proposed supplier, manufacturer, or distributor to ensure that requested sources and devices, where applicable, are compatible with and conform to the sealed source and device designations registered with NRC or an Agreement State. Licensees may not make any changes to the sealed source, device, or source/device combination that would alter the description or specifications from those indicated in the respective registration certificates, without obtaining NRC's prior permission in a license amendment. To ensure that applicants use sources and devices according to the registration certificates, they may want to get a copy of the certificate and review it or discuss it with the manufacturer.

Response from Applicant:

- For sealed materials:

  — Identify each radionuclide (element name and mass number) that will be used in each sealed source.

  — Provide the manufacturer's (distributor's) name and model number for each sealed source and, if applicable, device requested.

  — Confirm that the activity per source and maximum activity in each device will not exceed the maximum activity listed on the approved certificate of registration issued by NRC or by an Agreement State.

  — Confirm that each sealed source, device, and source/device combination is registered as an approved sealed source or device by NRC or an Agreement State.

A safety evaluation of sealed sources and devices is performed by NRC or an Agreement State before authorizing a manufacturer (or distributor) to distribute them to specific licensees. The safety evaluation is documented in a Sealed Source and Device (SSD) Registration Certificate. Information on SSD Registration Certificates may be obtained through the Registration Assistant by calling NRC's toll-free number, (800) 368-5642, extension 415-7231. Applicants must provide the manufacturer's name and model number for each requested sealed source and device (e.g., instrument calibrator) so that NRC can verify that each, when applicable, has been evaluated in an SSD Registration Certificate. See also NUREG-1556, Vol. 3.

- For unsealed tracer materials, including both volatile and nonvolatile materials (e.g., iodine-131, iodine-125, hydrogen-3):

  — Provide element name with mass number, chemical and/or physical form, and maximum requested possession limit

— Provide information for volatile materials, if known, on the anticipated rate of volatility or dispersion. This information may be obtained from the tracer material vendor, supplier, or manufacturer.

8.5.1   SEALED SOURCES AND DEVICES

Regulation: 10 CFR 30.32(g), 10 CFR 39.41.

Criteria: Any sealed source used for well logging that contains more than 3.7 MBq (100 microcuries) of byproduct or special nuclear material and is used downhole in well bores of gas wells, oil wells, or in mineral deposits, must satisfy one of the following criteria:

• Sealed sources that were manufactured before July 14, 1989, may use either the design and performance criteria from the United States of America Standards Institute (USASI) N5 10-1968 or the criteria specified in 10 CFR 39.41. The use of the USASI N5 10-1968 standard is based on an NRC Notice of Generic Exemption issued on July 25, 1989 (54 FR 30883), which has been included in NRC's final rule issued on April 17, 2000 (65 FR 20337). A copy of the referenced generic exemption letter is included in Appendix J as reference.

• Sealed sources are required to satisfy the requirements of 10 CFR 39.41.

The primary difference between the two standards is that the vibration requirement in 10 CFR 39.41 is not included in the USASI standard. This vibration test was included to ensure consistency between the United States standard and international standards.

Discussion: NRC or an Agreement State performs a safety evaluation of sealed sources before authorizing a manufacturer or distributor to distribute sources to specific licensees. The safety evaluation is documented in a Sealed Source and Device (SSD) Registration Certificate. Some examples of sealed sources used in well logging applications are shown in Figure 8.2.

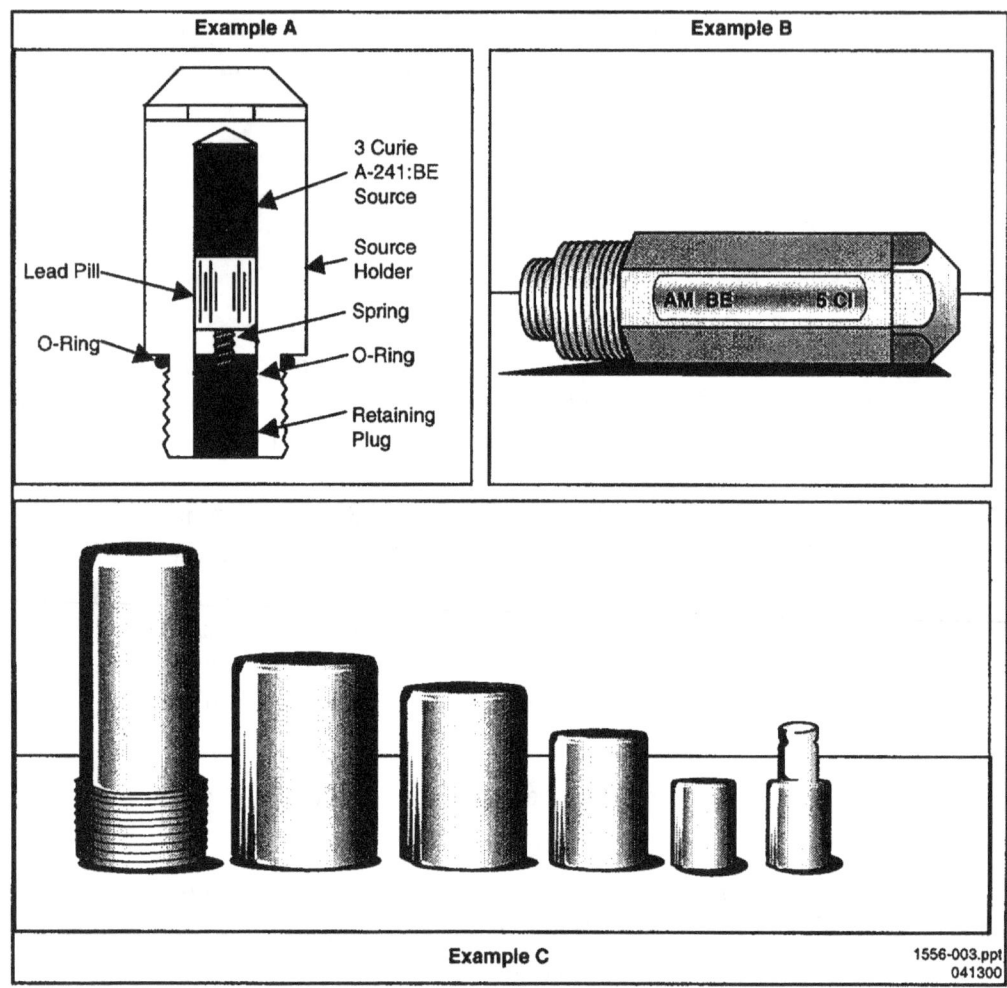

Figure 8.2 Examples of Sealed Sources Used In Well Logging Operations.

Applicants must provide the manufacturer's name and model number for each requested sealed source. This information is necessary to ensure that each sealed source requested in the application is evaluated and approved by NRC or an Agreement State, included in an SSD Registration Certificate, approved under the provisions granted by NRC in 10 CFR 39.41, or is identified on an NRC or Agreement State license and authorized for well logging. Applicants should consult with the proposed suppliers or vendors to ensure that sealed sources, and if applicable, devices, conform to information contained in SSD Registration Certificates. Licensees should ensure that their uses of sealed sources, and, if applicable, associated equipment are in accordance with Registration Certificates. Applicants may elect to obtain copies of applicable SSD Registration Certificates for future reference.

For sealed sources used for well logging applications, NRC licenses only authorize possession of individual sealed sources approved for well logging. To allow flexibility, licenses do not authorize specific sealed source/well logging tool combinations. Applicants should consult with the manufacturer of the sealed sources before using associated equipment, e.g., well logging tools, transport containers, handling tools, etc. Conferring with the vendor or manufacturer before use helps ensure that the associated equipment selected is compatible with sealed sources requested in the application.

Response from Applicant:

- Identify each sealed source with an activity greater than 3.7 MBq (100 microcuries) by the manufacturer's name, model number, and radionuclide (element name and mass number).

- Identify each energy compensation source with an activity less than or equal to 3.7 MBq (100 microcuries) by the manufacturer's name, model number, and radionuclide (element name and mass number).

- Confirm that each sealed source is registered as an approved sealed source by NRC or an Agreement State and will be possessed and used in accordance with the conditions specified in the Registration Certificate.

- Confirm that sealed sources not satisfying 10 CFR 39.41 performance requirements are approved by USASI N5 10-1968 standard for well logging (See Appendix J).

- Confirm that the activity per source and maximum activity in each device will not exceed the maximum activity listed on the approved certificate of registration issued by NRC or by an Agreement State.

- Provide the license number of an NRC or Agreement State license that approves a well logging source that is not included in an SSD registration certificate.

- Identify any sealed sources and/or corresponding devices not used in well logging that contain byproduct, special nuclear, or source material and specify the manufacturer's name, model number, and radionuclide (element name and mass number). An example of such a device is calibration devices used for survey instruments and pocket dosimeters, and sources used above ground for calibrating well logging tools.

- Identify the manufacturer's name and model number of depleted uranium sinker bars.

OR

- Complete the table in Appendix C to support the request for byproduct, source, or special nuclear material used in well logging operations and radioactive materials used for purposes other than well logging, e.g., radiation survey instrument calibrators.

Note: Information on SSD registration certificates is available electronically at <http://www.nrc.gov>; select the "Library" section. The current version of NUREG-1556, Vol. 3, "Consolidated Guidance About Materials Licenses: Applications for Sealed Source and Device Evaluation and Registration" provides specific information about the SSD registration process. This document is also available electronically at the above internet location, or for a paper copy of NUREG-1556, Vol. 3, see the Notice of Availability (on the inside front cover of this report). For individual copies of SSD registration sheets, an applicant may contact the Registration Assistant by calling NRC's toll free number, (800) 368-5642, and then asking for extension 415-7217.

8.5.2 UNSEALED (TRACER) RADIOACTIVE MATERIAL

Regulation: 10 CFR 30.32(i), 10 CFR 30.33, 10 CFR 30.72, 10 CFR 39.2, 10 CFR 39.13.

Criteria: An application for a license will be approved if the requirements of 10 CFR 30.33 and 10 CFR 39.13 are satisfied.

Discussion: Each authorized radioisotope tracer is listed on the NRC license by its element name, chemical and/or physical form, total possession limit, and the maximum amount of each radioisotope (identified by physical or chemical form) used in each type of tracer study requested. The following definitions are provided to clarify single and multiple well tracer operations addressed in this report.

- **Tracer Materials:** Radioactive isotopes in liquid, solid, or gas form that are injected into single well bores or underground reservoirs to monitor the movement of fluids or gases. Tracer studies involve a single well and require the use of an electronic well logging tool to detect the radioactive isotopes injected into the well.

- **Field Flood Studies or Enhanced Oil and Gas Recovery Studies:** Tracer studies involving multiple wells where one or more radioactive isotopes are injected and multiple oil or gas samples containing radioactive material are collected from each of the wells to determine the direction and rate of flow through the formation. Field flood tracer operations would not normally involve the use of an electronic well logging tool to detect the radioactive isotopes in the well.

- **Labeled Frac Sands:** Radioactive isotope(s) in liquid or solid forms that is(are) chemically bonded to glass and/or resin beads and injected into a single well in a density-controlled solution. Frac sand operations require the use of an electronic well logging tool to assess the amount of radioactive isotope(s) remaining in the underground reservoir formation.

See the sample license in Appendix E. Table 8.2 identifies the types of byproduct material used in tracer and field flood study applications covered by this report.

Table 8.2   Types of Radioactive Materials Used in Field Flood Studies and Single Well Tracer Operations

FIELD FLOOD OR ENHANCED OIL AND GAS RECOVERY STUDY APPLICATIONS TRACERS USED IN MULTIPLE WELLS	
Gas	H-3, Kr-85, C-14, I-131, Br-82
Liquid	H-3, C-14, Na-22, S-35, Ca-45, Co-60, Ni-63, Zn-65, Sr-85, Sc-46, Sr-90, Ag-110m, I-125, I-131, La-140, Ir-192
WELL LOGGING TRACER APPLICATIONS TRACERS USED IN A SINGLE WELL	
Gas	Br-82, I-131, I-125
Liquid	Fe-59, I-125, I-131, Sb-124, Au-198, Ag-110m
Labeled Frac Sand	Sc-46, Br-82, Ag-110m, Sb-124, Ir-192

Response from Applicant:

- For unsealed nonvolatile and volatile (e.g., iodine-125, iodine-131, hydrogen-3, bromine-82) tracer materials:

    — Provide the element name and mass number

    — Identify each chemical and/or physical form (e.g., liquid, gas, or labeled frac sands) requested for each type of tracer study

    — Specify the maximum amount of each radioisotope tracer material that will be possessed at any one time. Possession limits should also include any materials that may be stored as waste

    — Specify the maximum amount of each radioisotope tracer that you will use in each type of tracer study by its physical or chemical form. Identifying the forms as "any" is unacceptable.

- Provide an Emergency Plan (if required)

    — Emergency plans are not routinely required for tracer materials with half-lives of less than 120 days and for quantities authorized in well logging and tracer licenses. Applicants should refer to 10 CFR 30.72, Schedule C, to determine the quantities of radioactive material requiring an emergency plan for responding to a release.

8.5.3 FINANCIAL ASSURANCE AND RECORD KEEPING FOR DECOMMISSIONING

Regulations: 10 CFR 30.34(b), 10 CFR 30.35.

Criteria: Financial assurance is not required by most well logging or tracer licensees; however, each licensee is obligated to maintain, in an identified location, decommissioning records related to facilities where licensed material is used, stored, or dispatched. Decommissioning records described above are not required at temporary jobsites. Pursuant to 10 CFR 30.35(g), when terminating the license, licensees must transfer records important to decommissioning to either of the following:

- The new licensee before licensed activities are transferred or assigned according to 10 CFR 30.34(b)

- The appropriate NRC Regional office before the license is terminated.

Discussion: NRC regulations, when applicable, require the applicant to provide Certification of Financial Assurance (F/A) or a Decommissioning Funding Plan (DFP). This is to provide reasonable assurance that, after the technical and environmental components of decommissioning are carried out, unrestricted use of the facilities is possible at the termination of licensed activities. NRC's primary objective is to ensure that decommissioning will be carried out with minimum impact on the health and safety of the public, occupationally exposed individuals, and the environment (53 FR 24018). These requirements specify that a licensee either set aside funds for decommissioning activities or provide a guarantee through a third party that funds will be available (see Figure 8.3). Before a license is issued, applicants are required to submit F/A or a DFP when requesting authorization to possess any sealed or unsealed radioactive material with half life ($T_{1/2}$) greater than 120 days exceeding certain the limits. Criteria for determining whether an applicant must submit a DFP or has an option of submitting either a DFP or F/A are described in 10 CFR 30.35.

> There are two parts to this rule: financial assurance that applies to some licensees and record keeping that applies to *all* licensees.

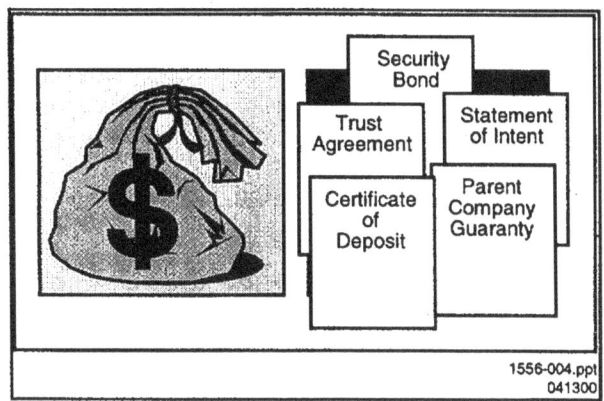

Figure 8.3 Methods of Certification of Financial Assurance for Decommissioning.

Table 8.3 provides a partial list of sealed and unsealed radioisotopes with T1/2 > 120 days with the corresponding limits in excess of which an F/A or a DFP is required. However, it is NRC's experience that most well logging, tracer, and field flood study licensees use only a few of these radioisotopes. The most frequently used radioisotopes requiring financial assurance in unsealed form are hydrogen-3, carbon-14, and silver-110 metastable, and for sealed sources, americium-241. Radioisotopes with T1/2 > 120 days are listed in Column 1. Column 2 lists the corresponding possession limits of radioisotopes requiring F/A. Column 3 lists the corresponding possession limits of unsealed radioisotopes requiring the submittal of a DFP. These limits apply when only one of these radioisotopes is possessed.

Applicants can use the data from Table 8.3 or the method given in Appendix I to determine if F/A is required and the amount that is required when more than one of these radioisotopes is requested.

Table 8.3   Commonly Used Licensed Materials Requiring Financial Assurance/Decommissioning Funding Plan

Column 1: Radioisotope	Column 2: Limit for F/A (millicuries*)	Column 3: Limit for DFP (millicuries*)
Unsealed Materials		
H-3	1,000	100,000
C-14	100	10,000
Ag-110m	1	100
Sealed Materials		
Am-241	100,000	N/A

*1 millicurie = 37 MBq

Regulatory Guide (RG) 3.66, "Standard Format and Content of Financial Assurance Mechanisms Required for Decommissioning Under 10 CFR Parts 30, 40, 70, and 72," dated June 1990, contains approved wording for each mechanism authorized by the regulation to guarantee or secure funds, except for the Statement of Intent for Government licensees.

Record Keeping

The requirements for maintaining records important to decommissioning, including the type of information required, are stated in 10 CFR 30.35(g). All licensees are required to maintain these records in an identified location until the site is released for unrestricted use (see Figure 8.4). In the event that the licensed activities are transferred to another person or entity, these records shall be transferred to the new licensee before transferring the licensed activities. The new licensee is responsible for maintaining these records until the license is terminated. When the license is terminated, these records shall be transferred to NRC.

Figure 8.4 Types of Records That Must Be Maintained for Decommissioning.

10 CFR 30.35(g), Requirements for Disposition of Records Important to Decommissioning

- Before licensed activities are transferred or assigned according to 10 CFR 30.34(b), transfer to the new licensee.

OR

- Before the license is terminated, transfer records to the appropriate NRC Regional office.

Response from Applicants: No response is needed from most applicants. If F/A or a DFP is required, submit the required documents described in Regulatory Guide 3.66.

Note: Licensees must maintain permanent records on locations where licensed materials are used or stored while the license is in force. These permanent records are important for making future determinations about the release of these locations for unrestricted use (e.g., before the license is terminated). Acceptable permanent records include sketches, written descriptions of specific locations where radioactive material is used or stored, and records of any leaking sealed sources, tracer material spills, contaminated waste storage areas, or other unusual occurrences involving the spread of contamination in or around the licensee's facilities or field stations. Permanent decommissioning records described above are not required for temporary job site locations.

References: See the Notice of Availability (on the inside front cover of this report) to obtain copies of RG 3.66 and Policy and Guidance Directive FC 90-2 (Rev. 1), "Standard Review Plan for Evaluating Compliance with Decommissioning Requirements," dated April 30, 1991.

8.6   ITEM 6: PURPOSE(S) FOR WHICH LICENSED MATERIAL WILL BE USED

Regulations: 10 CFR 30.33(a)(1), 10 CFR 39.13, 10 CFR 39.41, 10 CFR 39.45, 10 CFR 39.47, 10 CFR 39.49, 10 CFR 39.51, 10 CFR 39.63, 10 CFR 51.21.

Criteria: Radioisotopes and sealed sources requested in the application must be used for purposes authorized by the Atomic Energy Act of 1954, as amended. The licensee must specify the purpose for which each radioisotope or sealed source listed in Item 5 is to be used, as well as specifying the type of wells in which each type of material will be used (e.g., oil, gas, mineral, geophysical, etc.). In addition, the licensee should describe the type of mineral or geophysical logging to be conducted, e.g., coal, salt domes, etc. Sealed sources used in well logging devices should be used only for the purposes for which they were designed, in accordance with the manufacturer's written recommendations and instructions, as specified in an approved SSD Registration Certificate, and as authorized on an NRC or Agreement State license. The licensee shall specify the manufacturer and model number of each device.

Discussion: The applicant's request to use sealed sources and radioisotopes in well logging, tracer, and field flood studies should clearly specify the purpose for which each type of material will be used. Applicants should include a description that is sufficiently detailed to allow NRC to determine the potential for exposure to occupationally exposed individuals and/or members of the public.

Note: Traditionally, only Federal or State authorities have been authorized to conduct logging in potable water wells in fresh water aquifers. Approval to conduct these operations requires that applicants justify the need and to provide assurance that sealed sources, in case of accidental loss in a potable water zone, could be recovered.

Applicants requesting authorization to perform any of the hazardous operations listed below should clearly indicate their intent and provide specific instructions for conducting such activities in their operating and emergency procedures:

- Removing a sealed source from a source holder of a logging tool and maintenance on sealed sources or holders

- Using destructive techniques to remove a stuck sealed source from a source holder

- Opening, repairing, or modifying any sealed source

- Knowingly injecting licensed radioactive tracer material into a fresh water aquifer

- Using a sealed source in a well without a surface casing to protect fresh water aquifers.

Applicants may use the format given in Table 8.4 to provide the requested information.

Table 8.4   Sample Format for Providing Information About Requested Radioisotopes

Radioisotope	Chemical/Physical Form	Maximum Possession Limit	Proposed Use
Americium-241	Sealed neutron source (XYZ Inc., Model 10)	Not to exceed 5 curies per source	Oil, gas, and/or mineral logging.
Cesium-137	Sealed source (Okko Inc., Model 36)	Not to exceed 3 curies per source	Oil, gas, and/or mineral logging.
Hydrogen-3	Gas, titanium tritide neutron generator tube (Cols Inc., Model 3)	Not to exceed 3 curies per tube	Neutron activation logging in oil and gas wells in downhole accelerator

Radioisotope	Chemical/Physical Form	Maximum Possession Limit	Proposed Use
Iodine-131	Gas	100 millicuries total, not to exceed 20 millicuries per injection	Subsurface Tracer Operations
Iodine-131	Liquid	50 millicuries total, not to exceed 10 millicuries per injection	Subsurface Tracer Operations
Iridium-192	"Labeled" frac sand	200 millicuries total, not to exceed 15 millicuries per injection	Subsurface Tracer Operations
Cobalt-60	Metal wire	3 millicuries total, not to exceed 1 microcurie per individual unit	Pipe Joint Collar Markers, Subsidence Markers, Depth Determination
Silver-110m	Liquid	200 millicuries total, not to exceed 20 millicuries per injection	Field Flood Tracer Studies
Depleted Uranium	Sinker Bars	225 kilograms	Sinker Weights (Concentrated Mass)

If the material will be used in field flood studies where licensed material is intentionally released into the environment, an environmental assessment (EA) is required in accordance with 10 CFR 51.21. Revision 1, Supplement to Policy and Guidance Directive FC 84-20, "Impact of Revision of 10 CFR Part 51 on Materials License Actions," dated March 1994, provides criteria for determining when an EA is not needed.

Applicants should note that authorization granted by NRC to use licensed material in tracer or field flood studies does not relieve them of their responsibilities to comply with any other applicable Federal, State or local regulatory requirements.

Response from Applicant: List the specific use or purpose of each sealed source and/or radioisotope requested in the application.

8.7 ITEM 7: INDIVIDUAL(S) RESPONSIBLE FOR THE RADIATION SAFETY PROGRAM AND THEIR TRAINING AND EXPERIENCE

8.7.1 RADIATION SAFETY OFFICER (RSO)

Regulations: 10 CFR 30.33(a)(3).

Criteria: RSOs must have adequate training and experience.

Discussion: The person responsible for the radiation protection program is identified on the license as the RSO. The NRC believes the RSO is the key to overseeing and ensuring safe operation of the licensee's well logging, tracer, or field flood study program. The RSO needs independent authority to stop operations that he or she considers unsafe. He or she must have sufficient time and commitment from management to fulfill certain duties and responsibilities to ensure that radioactive materials are used in a safe manner. The RSO may delegate certain day-to-day tasks of the radiation protection program to other responsible individuals without delegating his or her responsibilities of the radiation safety program. For example, a large well logging firm with multiple field stations and/or temporary job sites may appoint individuals designated as "site RSOs" who assist the RSO and are responsible for the day-to-day activities at the field stations and/or temporary job sites.

Typical RSO duties are illustrated in Figure 8.5 and Appendix K. NRC requires the name of the RSO on the license to ensure that licensee management has always identified a responsible, qualified person and that the named individual knows of his or her designation as RSO. Provide NRC with a copy of an organizational chart showing the RSO (and other designated responsible individuals) to demonstrate that he or she has sufficient independence and direct communication with responsible management officials. Also, show in the organizational chart the position of the individual who signs the application in Item 13 of the NRC Form 313.

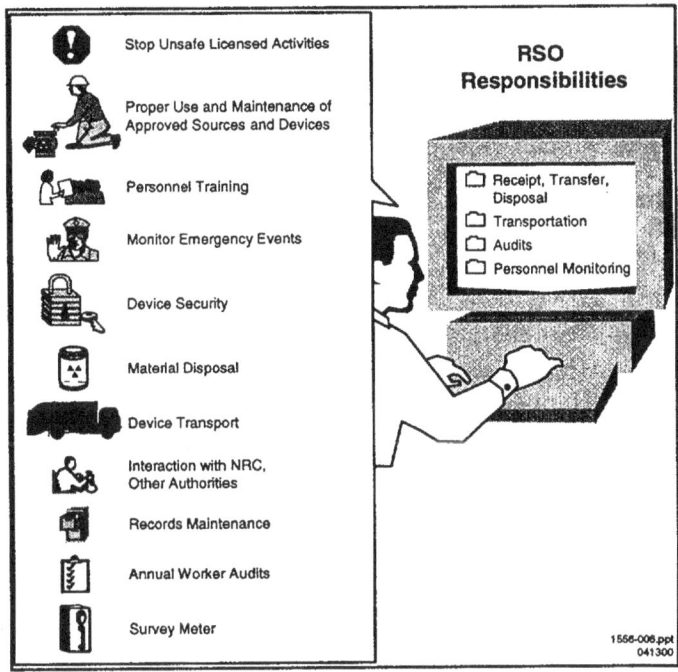

Figure 8.5 RSO Responsibilities - Typical duties and responsibilities of the RSO.

To be considered eligible for the RSO position, the applicant must submit for review the specific training and experience of the proposed RSO and detail his or her duties and responsibilities. The proposed RSO should have had a minimum of 1 year of actual experience as a logging supervisor. The RSO is expected to coordinate the safe use of licensed materials and to ensure compliance with the applicable requirements of the Code of Federal Regulations (e.g., Parts 19, 20, 21, 30, 39, etc.). The RSO should possess a thorough knowledge of management policies, company administrative and operating procedures, and safety procedures related to protection against radiation exposures.

Response from Applicant: Provide the following:

- The name of the proposed RSO who will be responsible for ensuring that the licensee's radiation safety program is implemented in accordance with approved procedures

AND

- Demonstrate that the RSO has sufficient independence and direct communication with responsible management officials by providing a copy of an organization chart with positions demonstrating day-to-day oversight of the radiation safety activities

AND EITHER

- The specific training and experience of the RSO

OR

- Alternative information demonstrating that the proposed RSO is qualified by training and experience (e.g., Board Certification by the American Board of Health Physicists; completion of a bachelor's and/or master's degree in the sciences with at least one year of experience in the conduct of a radiation safety program of comparable size and scope)

- Formal training in the establishment and maintenance of a radiation protection program

OR

- Alternative information demonstrating that the proposed RSO is qualified by training and experience, e.g., listed by name as an authorized user or the RSO on an NRC or Agreement State license that requires a radiation safety program of comparable size and scope.

Note: It is important to notify NRC and obtain a license amendment prior to making changes in the designation of the RSO responsible for the radiation safety program.

8.8 ITEM 8: TRAINING FOR LOGGING SUPERVISORS AND LOGGING ASSISTANTS

Regulations: 10 CFR 19.11, 10 CFR 19.12, 10 CFR 19.13, 10 CFR 30.7, 10 CFR 30.9, 10 CFR 30.10, 10 CFR 30.33, 10 CFR 39.13, 10 CFR 39.61.

Criteria: Well logging supervisors and well logging assistants must have adequate training and experience as outlined in 10 CFR 19.12, 10 CFR 30.33(a)(3), and 10 CFR 39.61. **Although persons engaged in field flood studies operations are not specifically addressed in 10 CFR Part 39, NRC staff has historically accepted classroom training for tracer studies to be an appropriate guide for individuals engaged in field flood studies.**

Discussion:

- A logging supervisor is a person who performs or personally supervises well logging operations, tracer/field flood study applications and is responsible for ensuring compliance with NRC regulations and the safe use of radioactive materials.

- A logging assistant is an individual, who under the ***direct supervision and in the physical presence of the logging supervisor*** uses well logging equipment (sealed sources containing byproduct material, related handling tools, unsealed sources of byproduct material, well logging devices, and radiation survey instruments) in performing well logging operations.

Didactic training and testing requirements, performance requirements, annual refresher training, and annual audit requirements for logging supervisors and logging assistants are outlined in 10 CFR 39 61.

Refer to Appendix L as an aid in determining the specific training requirements for logging supervisors, logging assistants, and individuals authorized to conduct field flood study/tracer applications. The applicant must submit a description of its training program for logging supervisors, logging assistants, and/or individuals authorized to conduct field flood study applications.

Because 10 CFR Part 39 contains different requirements for logging supervisors and logging assistants, applicants must include training programs for each category. When describing the training programs for these positions, include the sequence of events from the time of hiring through the designation of individuals as logging supervisors or logging assistants. Experienced logging supervisors who have worked for another well logging, tracer, or field flood study licensee should receive formal instruction similar to that given to prospective logging assistants.

Instructors who provide classroom training to individuals in the principles of radiation and radiation safety should have knowledge and understanding of these principles beyond those obtainable in a course similar to the one given to prospective logging supervisors. Individuals who provide instruction in the hands-on use of well logging and handling equipment should be qualified logging supervisors with at least 1 year of experience in performing well logging operations, or should possess a thorough understanding of the operation of well logging and handling equipment (e.g., a manufacturer's service representative).

An internal inspection program (audit) of the job performance of each logging supervisor and logging assistant ensures that the Commission's regulations, license requirements, and the licensee's operating and emergency procedures are followed. The audit must include observation of the performance of each logging supervisor and logging assistant during an actual well logging operation at intervals not to exceed 12 months. If a logging supervisor or logging assistant has not participated in a well logging operation for more than 12 months since the last inspection, the individual must be inspected the first time he or she engages in well logging operations.

Response from Applicant:

- Submit an outline of the training to be given to prospective logging supervisors and logging assistants. Submit your procedures for experienced logging supervisors who have worked for another licensee.

- Provide a copy of a typical examination and the correct answers to the examination questions. Indicate the passing grade.

- Specify the qualifications of your instructors in radiation safety principles and describe their experience with well logging activities. If training will be conducted by someone outside the applicant's organization, identify the course by title, provide the name, address, and telephone number of the company providing the training, and provide a course outline (if available).

- Describe the field (practical) examination that will be given to prospective logging supervisors and logging assistants. The NRC suggests using the checklist in Appendix M as a source of potential areas to review during the field examination.

- Describe the annual refresher training program, including topics to be covered and how the training will be conducted.

- Submit a description of your program for inspecting the job performance of each well logging supervisor or logging assistant at intervals not to exceed 12 months, as described in 10 CFR 39.13.

8.9   ITEM 9: FACILITIES AND EQUIPMENT

Regulations: 10 CFR 20.1406, 10 CFR 20.1101(b), 10 CFR 20.1703, 10 CFR 30.33(a)(2), 10 CFR 30.35(g), 10 CFR 39.31(b)(1), 10 CFR 39.45(a), 10 CFR 39.71, 10 CFR 40.32(c), 10 CFR 70.23(a)(3).

Criteria: Facilities and equipment must be adequate to protect health, minimize danger to life or property, minimize the possibility of contamination, and keep exposure to occupationally exposed workers and the public ALARA.

Discussion: Applicants must demonstrate that proposed facilities and equipment provide adequate storage capabilities, ensure that appropriate shielding is available to protect the health and safety of the public and employees, keep exposures to radiation and radioactive materials ALARA, and minimize the possibility of contamination from the uses, types, and quantities of radioactive materials requested.

Licensed materials located in an unrestricted area and not in storage must be under the constant surveillance and immediate control of the licensee. Areas where material is used or stored, including below ground bunker storage areas, should (1) be accessible only by authorized persons; and (2) secured or locked when an authorized person is not physically present. Use or storage areas cannot be considered restricted areas for purposes of radiation safety if accessible by unauthorized persons.

Applicants may delay completing facilities and acquiring equipment until after the application review is completed. Delaying the acquisition will allow for changes, if any, needed as a result of the application review. This delay will also ensure the adequacy of proposed facilities and equipment before the applicant makes a significant financial commitment. In all cases, the applicant cannot possess or use licensed material until after the facilities are approved, equipment is procured, and the license is issued.

Response from Applicant:

- Submit a drawing or sketch of the proposed facility identifying areas where radioactive materials, including radioactive wastes, will be used or stored.

- Show in drawings, where applicable, adjacent buildings, boundary lines, security fences, and lockable storage areas.

- Illustrate area(s) where explosive, flammable, or other hazardous materials may be stored.

- Show in the drawings the relationship and distance between restricted areas and adjacent unrestricted areas.

- Specify in the drawings shielding materials (concrete, lead, etc.) and means for securing radioactive materials from unauthorized removal.

- Draw to an indicated scale, or include dimensions on each drawing or sketch.

- Submit a drawing or sketch of the proposed tracer material storage facilities including rooms, buildings, below ground bunker storage areas, or containers used for storage of both tracer and tracer waste materials, if appropriate. Specify the types and amount of shielding materials (concrete, lead, etc.) and means for securing tracer materials from unauthorized removal.

- Describe protective clothing (such as rubber gloves, coveralls, respirators, and face shields), auxiliary shielding, absorbent materials, injection equipment, secondary containers for waste water storage for decontamination purposes, plastic bags for storing contaminated items, etc., that will be available at well sites when using tracer materials.

- Describe proposed laundry facilities, if applicable, used for contaminated protective clothing, and specify how the contaminated waste water from the laundry machines or sinks is disposed. Operating and emergency procedures should address decontamination of the laundry area and equipment.

- Describe proposed decontamination facilities for trucks, tracer injection tools, or other equipment contaminated by tracer materials, if applicable. Specify how the contaminated waste water for these decontamination facilities is disposed. Operating and emergency procedures should address decontamination of these types of equipment and facilities.

- Describe, if applicable, equipment for "repackaging" gaseous, volatile, or finely divided tracer material. Most tracer users do not repackage materials and acquire their injections in precalibrated amounts or "ready to use" forms. However, should an applicant request the ability to repackage tracer, volatile, or finely divided material, consider the following equipment when repackaging tracer materials: sinks, trays with absorbent material, glove boxes, fume hoods with charcoal filtration, filtered exhaust, special handling equipment including special tools, rubber gloves, etc.

> 10 CFR 20.2003 authorizes the disposal of readily soluble radioactive materials via the sanitary sewage. Sanitary sewage does not include sewage treatment facilities, septic tanks, and leach fields owned or operated by a licensee.

8.10   ITEM 10: RADIATION SAFETY PROGRAM

A radiation safety program must be established and submitted to the NRC as part of the application. The program must be commensurate with the scope and extent of activities for the use of licensed materials in well logging, tracer, and field flood study operations. Each applicant must develop, document, and implement a radiation protection program containing the following elements:

- Development and implementation of an ALARA program

- Description of equipment and facilities adequate to protect personnel, the public and the environment

- Confirmation that licensed activities are conducted only by individuals qualified by training and experience

- Development and maintenance of written operating and emergency procedures

- Implementation of an audit program to inspect the job performance of well logging supervisors and assistants

- Description of organization structure and individuals responsible for ensuring day-to-day oversight of the radiation safety program

- Establishment and management of a radiation safety and decommissioning records system.

Discussion: Individual components of a radiation safety program are addressed in the topics found in this NUREG. Some topics will not require the applicant to submit information as part of an application, but simply provide the applicant with guidance to comply with a specific NRC requirement.

Applicants who plan to conduct well logging operations using sealed sources, tracer materials or tracer materials in field flood study operations are required to submit for NRC approval their Operating and Emergency procedures or, optionally, to provide either an outline or summary of each procedure that includes the important radiation safety aspects of each individual procedure. Radiation safety programs including tracer materials must assure that they address these additional concerns:

- Methods or procedures for preventing the release of contaminated material, equipment or vehicles to unrestricted use from tracer or field flood study operations

- Radiation safety procedures and the well logging supervisors' responsibilities unique to tracer and field flood study operations

- Tracer and field flood study equipment, techniques, and corresponding radiation safety procedures associated with use of tracer materials.

Note: Appendix F includes a description of procedures for using tracer materials in field flood study operations.

Response from Applicant: The applicant is required to establish and submit its radiation protection program. Each bulleted item listed above should be addressed.

8.10.1 WELL OWNER/OPERATOR AGREEMENTS

Regulations: 10 CFR 39.15(a), 10 CFR 39.15(d), 10 CFR 39.69(c), 10 CFR 39.77(c), 10 CFR 39.77(d).

Criteria: Well logging conducted with a sealed source shall only be performed if a written agreement with the employing well owner or operator is executed prior to the start of well logging operations.

Discussion: Well logging operations conducted using a sealed source are performed only after a written agreement is executed with the employing well owner or operator. Written agreements must identify a responsible party for ensuring that the following steps will be taken if a source becomes lodged in a hole:

- A reasonable effort will be made to recover the source

- A person will not attempt to recover a lodged sealed source in a manner that, in the licensee's opinion, could result in its rupture

- During efforts to recover a sealed source, a licensee must continuously monitor the circulating fluids in the well bore, as required in 10 CFR 39.69(c)

- Contaminated equipment, personnel, or environment must be decontaminated prior to release

- If a sealed source is classified by the licensee as irretrievable after reasonable efforts at recovery have been expended, the following must be implemented within 30 days, as shown in Figure 8.6:

  — Source must be immobilized and sealed in place with a cement plug

  — Provide a means to prevent inadvertent intrusion on the source, unless the source is not accessible to any subsequent drilling operations

  — Install a permanent identification plaque at the surface of the well, unless mounting of a plaque is not practical. Figure 8.7 provides a diagram of a permanent identification plaque, describing the information that should be included on the plaque.

— Notify the appropriate NRC Regional Office by telephone of the circumstances that resulted in the inability to retrieve the source and obtain approval to implement abandonment procedures.

• Send a copy of the abandonment report within 30 days of the abandonment of the sealed source, to the appropriate NRC Regional Office and each appropriate State or Federal Agency that issued permits or otherwise approved of the drilling operation. The abandonment report must contain all the information outlined in 10 CFR 39.77(d). Refer to Appendix Q for additional guidance.

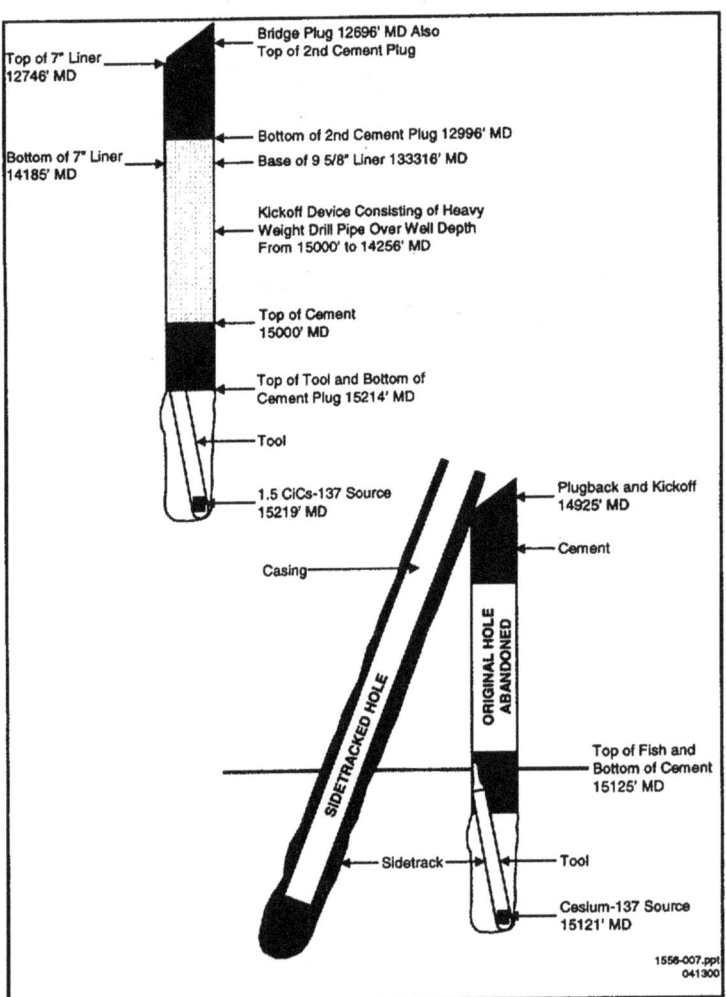

Figure 8.6 Features of a Typical Source Abandonment.

The NRC is aware that in some circumstances, such as high well pressures that could lead to fires or explosions, the delay required to obtain NRC approval to abandon the well may introduce an immediate threat. ***Under such exigent circumstances, immediate abandonment, without prior NRC approval, is authorized if a delay could cause an immediate threat to public health and safety.*** The NRC would then be notified as soon as possible after the abandonment. See 10 CFR 39.77(c)(1) and (d).

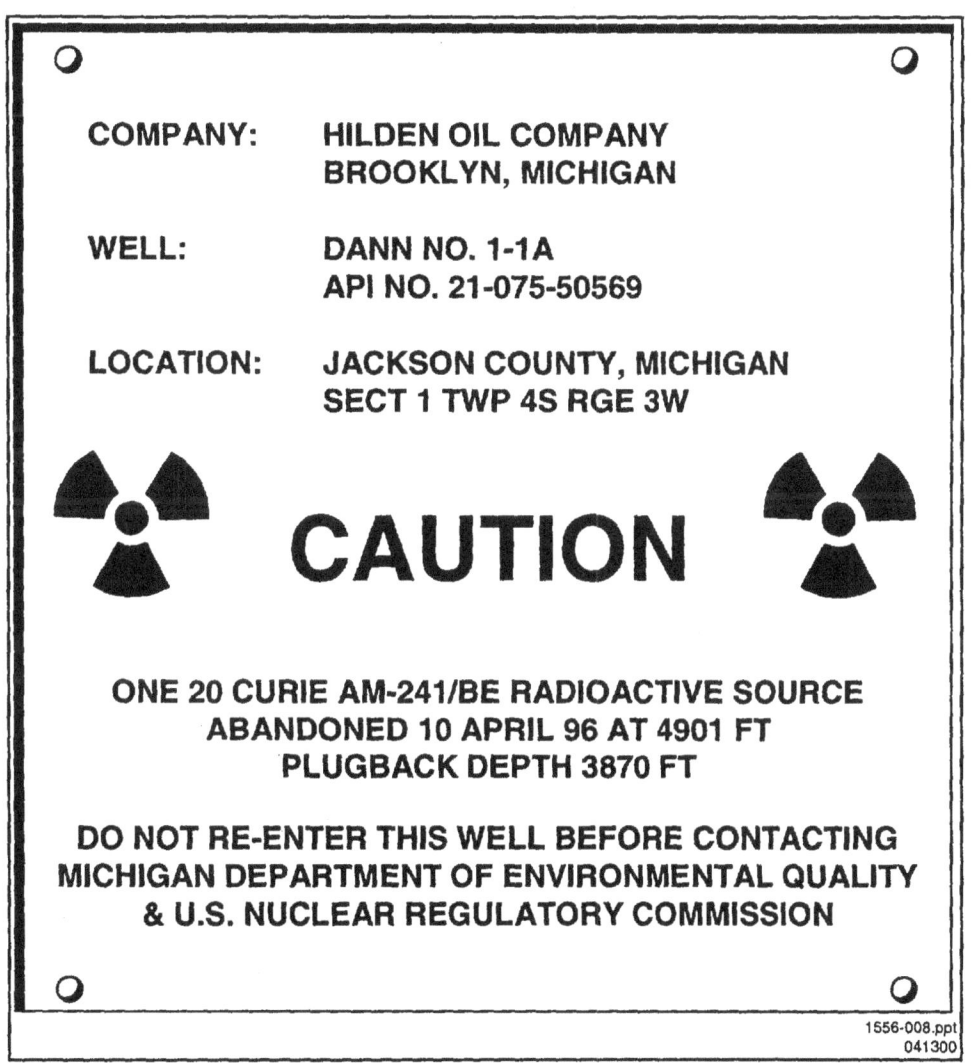

Figure 8.7 Permanent Identification Plaque.

Note: A written agreement is not required if the licensee and well owner or operator are part of the same corporate structure or otherwise similarly affiliated. However, all other requirements must still be met.

- If the requirement for a written agreement does not apply to you, then you should include a statement in your application that you will only log holes where the well owner or operator is part of your corporate structure or otherwise similarly affiliated, and you should describe the corporate affiliation.

Response from Applicant: Provide the following:

A statement that: "We will obtain a written agreement prior to well logging with a sealed source that meets the requirements specified in 10 CFR 39.15."

8.10.2 RADIATION SAFETY PROGRAM AUDIT

Regulations: 10 CFR 20.1101, 10 CFR 20.2102.

Criteria: Licensees must review the content and implementation of their radiation protection programs annually to ensure the following:

- Compliance with NRC and DOT regulations (as applicable), and the terms and conditions of the license

- Occupational doses and doses to members of the public are as low as reasonably achievable (ALARA) (10 CFR 20.1101)

- Records of audits and other reviews of program content and implementation are maintained for 3 years.

Discussion: Licensees are encouraged to implement as part of the radiation safety program a self-assessment and corrective action tracking program. Assessments necessary to ensure safe operations should result in a continuous process to self-identify violations, implement immediate corrective action when required, and track to completion and close-out of self-identified violations. NRC's enforcement policy is designed to encourage and to give credit to licensees for self-identifying violations and for taking immediate corrective actions. NRC policy allows licensees with a good regulatory performance, as shown by a licensee's inspection history, to be inspected less frequently than licensees where NRC staff identifies significant violation(s) during an inspection. Although the annual ALARA audit required by 10 CFR 20.1101(b) is an important cornerstone of the radiation safety program, NRC encourages applicants/licensees to develop and implement an ongoing audit program and corresponding corrective action tracking program.

Appendix G contains a suggested annual audit program that is specific to well logging and tracer operations and is acceptable to NRC. All areas indicated in Appendix G may not be applicable to every licensee and may not need to be addressed during each audit.

Response from Applicant: The applicant is not required to, and should not, submit its radiation safety program audit (ALARA) to the NRC for review during the licensing phase. The applicant's program for reviewing the content and implementation of its radiation safety program will be examined during inspection.

References: The current version of NUREG-1600 is available electronically at <http://www.nrc.gov/OE>. INs are available in the "Reference Library" on NRC's Home Page at <http://www.nrc.gov>. For hard copies of NUREG-1600, IN 96-28, and MC 87113, Appendix A, "Well Logging Inspection Field Notes," see the Notice of Availability (on the inside front cover of this report).

8.10.3 RADIATION MONITORING INSTRUMENTS

Regulations: 10 CFR 20.1501, 10 CFR 20.2103(a), 10 CFR 30.33(a)(2), 10 CFR 39.33.

Criteria: Licensees must possess radiation monitoring instruments that are necessary to protect health and minimize danger to life or property. Instruments used for quantitative radiation measurements must be calibrated for the radiation that is measured at least every 6 months. For the purposes of this document, survey instruments are defined as any device used to measure the radiological conditions at a licensed facility, field station, or temporary job site.

Discussion: For well logging and tracer operations, instruments must be capable of measuring 0.1 milliroentgen (2.58 X 10-8 C/kg) per hour through at least 50 milliroentgen (1.29 X 10-5 C/Kg) per hour. Licensees shall possess operable and calibrated radiation detection/measurement instruments to perform the following, as necessary:

- Package surveys

- Vehicle surveys

- Tracer material contamination surveys of equipment, vehicles, personnel and sites

- Prescreening of sealed source leak tests

- Unrestricted area dose rate measurements.

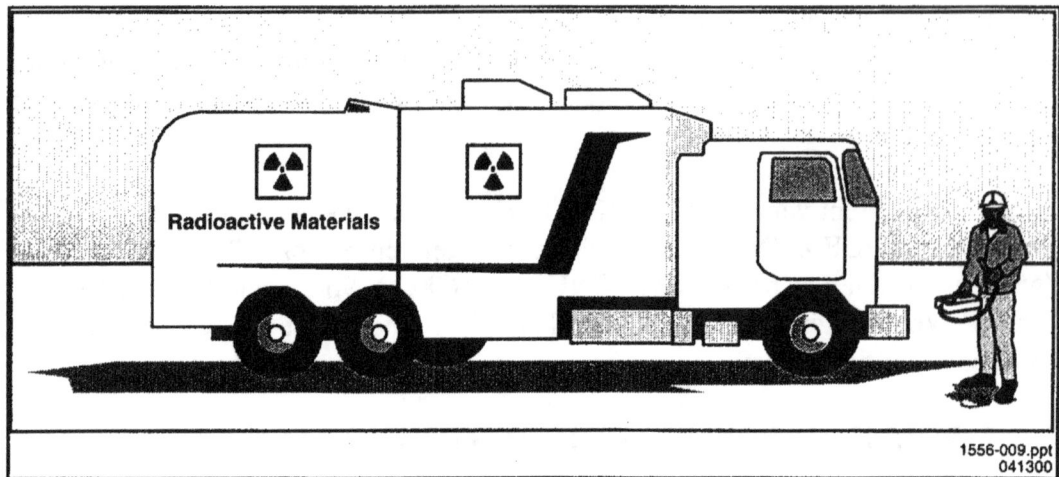

Figure 8.8 Types of Surveys. *There are many different types of surveys performed by well logging, tracer, and field flood studies licensees.*

The choice of instrument should be appropriate for the type of radiation to be measured, and for the type of measurement to be taken (count rate, dose rate, etc.).

Applications should include descriptions of the instrumentation available for use and instrumentation that applicants intend to purchase prior to starting licensed activities. The description should include type of instrument and probe, and the instrument's intended purpose.

NRC requires that calibrations be performed by the instrument manufacturer or a person specifically authorized by NRC or an Agreement State, unless the applicant specifically requests this authorization. Applicants seeking authorization to perform survey instrument calibrations shall submit procedures for review. Appendix N provides information about instrument specifications and model calibration procedures.

Response from Applicant: Provide one of the following:

- A description of the instrumentation (as described above) that will be used to perform required surveys and a statement that: "We will use instruments that meet the radiation monitoring instrument specifications published in Appendix N to NUREG-1556, Vol. 14, 'Program-Specific Guidance About Well Logging, Tracer and Field Flood Studies,' dated June 2000. We reserve the right to upgrade our survey instruments as necessary."

OR

- A description of the instrumentation (as described above) that will be used to perform required surveys and a statement that: "We will use instruments that meet the radiation monitoring instrument specifications published in Appendix N to NUREG-1556, Vol. 14, 'Program-Specific Guidance About Well Logging, Tracer and Field Flood Studies,' dated June 2000. Additionally, we will implement the model survey meter calibration program published in

Appendix N to NUREG-1556, Vol. 14, 'Program-Specific Guidance About Well Logging, Tracer and Field Flood Studies,' dated June 2000. We reserve the right to upgrade our survey instruments as necessary."

OR

- A description of alternative equipment and/or procedures for ensuring that appropriate radiation monitoring equipment will be used during licensed activities and that proper calibration and calibration frequency of survey equipment will be performed. Further, the statement "We reserve the right to upgrade our survey instruments as necessary" should be added to the response.

Note: Alternative responses will be reviewed using the criteria listed above.

8.10.4 MATERIAL RECEIPT AND ACCOUNTABILITY

Regulations: 10 CFR 20.1801, 10 CFR 20.1802, 20.1906, 10 CFR 30.34(e), 10 CFR 30.35(g),10 CFR 30.41, 10 CFR 30.51(g)(2), 10 CFR 39.37.

Criteria: Licensees with licensed material must do the following:

- Maintain records of receipt, transfer, and disposal of licensed materials

- Conduct physical inventories of licensed materials at least every 6 months to account for all sealed sources, tracer materials, and depleted uranium

- Maintain inventory records 3 years from the date of the inventory.

Discussion: Licensed materials must be tracked from the time of receipt to disposal in order to ensure accountability, identify when licensed material is lost, stolen, or misplaced, and to ensure that possession limits listed on the license are not exceeded. Physical inventories include locating, verifying the physical presence, and/or accounting for materials by the use of material receipt and transfer records.

Inventory records must contain the following types of information:

- Quantity and kind of licensed material including sealed sources, tracer material on hand (including waste), and depleted uranium in sinker bars

- Location of each sealed source

- Date the inventory occurred

- Name of individual performing the inventory.

Note: Physical inventory records may be combined with leak test records.

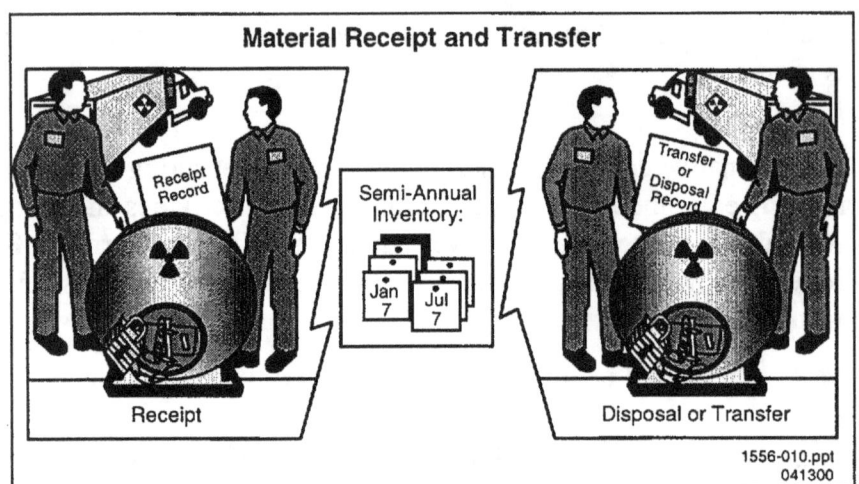

Figure 8.9 Material Receipt and Accountability. *Licensees must maintain records of receipt and disposal and conduct physical inventories at intervals not to exceed 6 months.*

Response from Applicant: Provide the following:

A statement that: "Physical inventories will be conducted and documented at least every 6 months to account for all licensed material, including byproduct, tracer, and depleted uranium received and possessed under the license."

8.10.5 OCCUPATIONAL DOSIMETRY

Regulations: 10 CFR 20.1201, 10 CFR 20.1207, 10 CFR 20.1208, 10 CFR 20.1501, 10 CFR 20.1502, 10 CFR 39.65.

Criteria: According to 10 CFR 39.65, logging supervisors and logging assistants must wear either film badges or thermoluminescent dosimeters (TLDs) during the handling or use of licensed radioactive material. This requirement applies to personnel using dosimeters for whole body measurements. Although not included in 10 CFR 39.65, some Agreement States have authorized Optically Stimulated Luminescence (OSL) dosimetry devices approved by the National Voluntary Laboratory Accreditation Program (NVLAP). NRC is currently in the process of amending its regulations to authorize the use of OSL dosimetry devices. *However, if a licensee wants to use OSL dosimetry until NRC's regulations are changed, it is necessary for an applicant to specifically request authorization to use OSL dosimetry.* Licensees must provide to employees, either a film or TLD that is processed by an accredited entity under the NVLAP operated by the National Institute of Standards and Technology (NIST).

Appendix O provides guidance for determining if individuals other than the RSO, logging supervisors, or logging assistants require dosimetry.

Bioassay services required in a license must be provided to individuals using tracer materials in subsurface studies if required by the license.

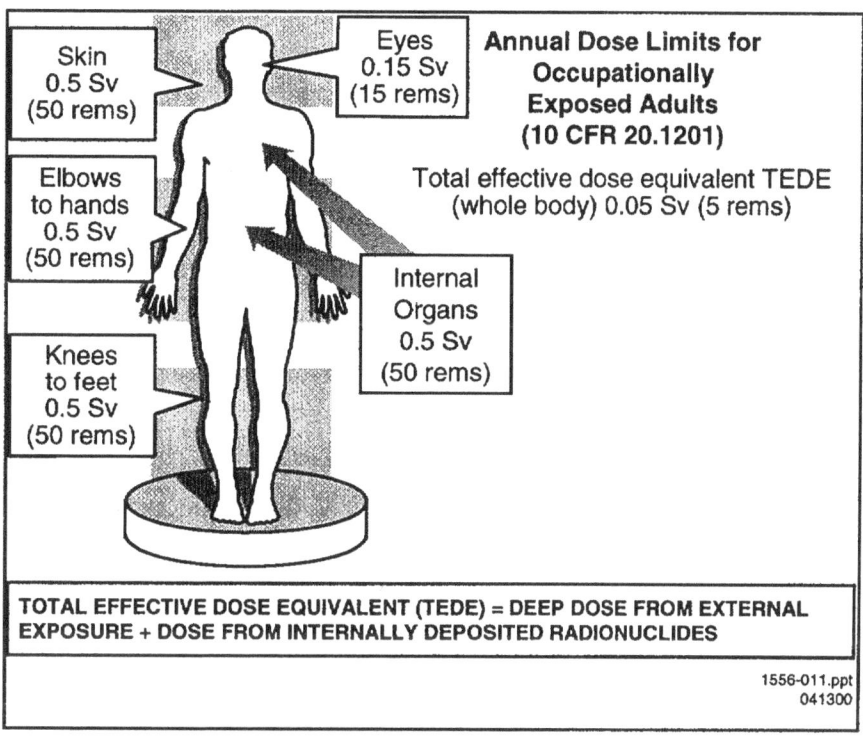

Figure 8.10 Annual Dose Limits for Occupationally Exposed Adults.

Discussion: The licensee may not permit any individual to act as a logging supervisor or logging assistant unless, at all times during the handling of licensed radioactive material, each individual wears on the trunk of the body a NVLAP-approved film badge, TLD, or OSL/personnel dosimeter (if specifically approved by NRC) that is sensitive to the type of radiation(s) to which the individual is exposed. If neutron sources are to be used, a commitment to provide neutron-sensitive dosimetry devices is required. Film badges must be replaced at intervals not to exceed 1 month, and TLDs or OSL must be replaced at intervals not to exceed 3 months.

For purposes of internal dosimetry, bioassays are required when individuals work with volatile radioactive material in the quantities, chemical and physical forms, and activities that make it likely that the radionuclide will be ingested, inhaled, or absorbed resulting in an intake in excess of 10% of the applicable annual limit on intakes (ALIs) in table 1, Columns 1 and 2, of Appendix B to 10 CFR Part 20. One ALI results in a CEDE of 5 rems or a CDE of 50 rems.

When using individually packaged "ready to use" quantities of iodine-131 tracer materials in well logging operations, bioassays are required for individuals using more than 50 millicuries at any one time, or using a total of 50 millicuries within any 5-day period. Guidance on bioassay programs for iodine-131, including the levels and types of handling for which bioassays are indicated, is provided in Regulatory Guide 8.20, "Applications of Bioassay for iodine-125 and iodine-131." Copies may be obtained from NRC's Regional Offices or at locations identified on the inside cover of the report in the Notice of Availability.

Bioassay services are available and provided by local hospitals, universities, or other vendors specifically approved to provide such services.

Response from Applicant:

Provide the following:

- A statement that the required film badge, TLD, or OSL dosimeter, processed and evaluated by a NVLAP-accredited entity and exchanged at the approved frequency, will be worn by well logging personnel.

To obtain a copy of the NIST Publication 810, "National Voluntary Laboratory Accreditation Program, 1997 Directory," contact the Superintendent of Documents, U.S. Government Printing Office, Washington, DC 20402-9225. (For information on the program, call NIST at 301-975-3679.) Also, NVLAP maintains a directory of accredited laboratories on the Internet (updated quarterly). The URL for NVLAP's home page on the Internet is <http://ts.nist.gov/nvlap>.

AND/OR
- Provide a bioassay program when using unsealed radioactive tracer materials. If an applicant elects to provide a bioassay program that is less conservative than recommended in Regulatory Guide 8.20, its rationale should be stated.

OR
- In lieu of providing a bioassay program, applicants may provide a commitment that they will not allow any individual to use more than 50 millicuries of iodine-131 at any one time or in any 5-day period at field stations or at temporary job sites. However, if an applicant plans to use an excess of the amounts described above or requests permission to repackage or process iodine-131 tracer materials at field stations, it is necessary to describe in detail the bioassay program. Bioassay programs should include what the applicant considers an acceptable interval or schedule for conducting bioassays, identify action levels or guidelines, and describe specific actions to be taken when action levels are exceeded. Because of the complex nature of bioassay and corresponding data analysis, it is acceptable for applicants to make reference to the procedures in NRC guidance documents.

OR

- Contract with an outside group for bioassay services. Provide a commitment that each vendor is licensed or otherwise authorized by NRC or an Agreement State to provide required bioassay services.

8.10.6 PUBLIC DOSE

Regulations: 10 CFR 1301, 10 CFR 20.1302, 10 CFR 20.1801, 10 CFR 20.1802, 10 CFR 20.2107.

Criteria: Licensees must do the following:

- Ensure that licensed material will be used, transported, stored, and disposed of in such a way that members of the public will not receive more than 1 mSv (100 mrem) in one year, and the dose in any unrestricted area will not exceed 0.02 mSv (2 mrem) in any one hour, from licensed operations

- Control and maintain constant surveillance of licensed material when in use and not in storage

- Secure stored licensed material from access, removal, or use by unauthorized personnel.

Discussion: Members of the public include persons who work in or may occupy locations where licensed material is used or stored. Employees whose assigned duties do not include the use of licensed material and work in the vicinity where it is used or stored are also included as members of the public. Public dose is controlled, in part, by ensuring that licensed material is secured (e.g., located in a locked area) to prevent unauthorized access or use. Well logging sealed sources and tracer materials are usually restricted by controlling access to the keys needed to gain access to storage locations, including downhole storage bunkers.

Public dose is also affected by the choice of storage and use locations at the field stations and at temporary job sites. Licensed material must be located so that the resulting public dose in an unrestricted area (e.g., an office or the exterior surface of an outside wall) does not exceed 1 mSv (100 mrem) in a year or 0.02 mSv (2 mrem) in any one hour. Applicants should use the concepts of controlling time, distance, and shielding when choosing storage and use locations. Decreasing the time that an individual is exposed, increasing the distance from the radioactive material, and adding shielding that is appropriate for the specific type of radiation (e.g., brick, concrete, lead, hydrogenous materials, etc.) will reduce the radiation exposure.

Information provided by the manufacturer or vendor on anticipated radiation levels of sealed sources and tracer materials, both inside their respective transport containers and outside the transport container at given distances, is the type of information needed to make public dose calculations. Licensees may assess radiation levels located in adjacent areas to radioactive material either by making calculations or by using a combination of direct measurements and calculations. After obtaining anticipated radiation levels or by making direct radiation measurements using an appropriate survey instrument, an applicant can use the "inverse square" law to evaluate the effect on the public and use this information to determine operating and emergency procedures for using radioactive materials. See Appendix P for an example demonstrating that individual members of the public will not receive doses exceeding the allowable public limits.

If, after making an initial public dose evaluation, a licensee changes the conditions used for the evaluation (e.g., relocates radioactive material within a designated storage area, increases the amount of radioactive materials in storage, changes the frequency radioactive material is in use, or changes the occupancy of adjacent areas), the licensee must perform a new evaluation to ensure that the public dose limits are not exceeded and take corrective action, if required.

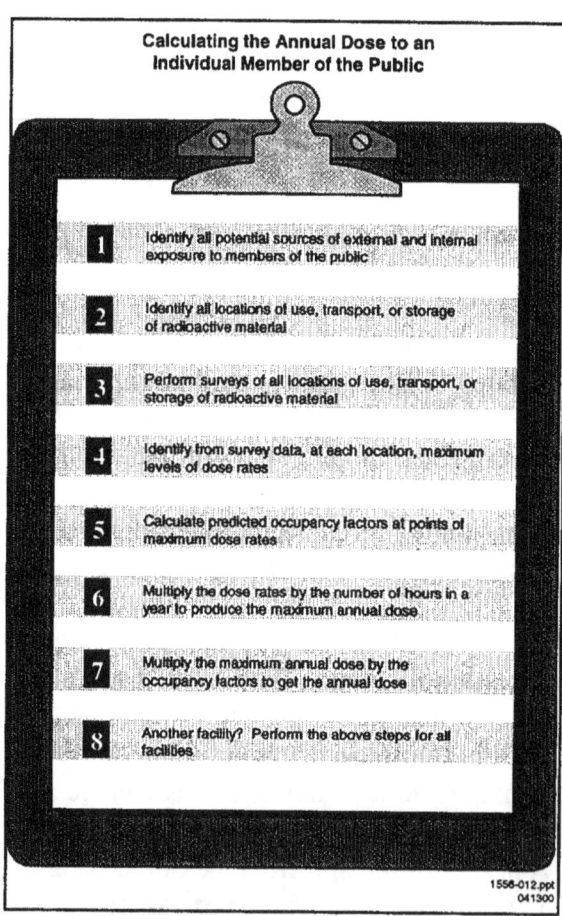

Figure 8.11 Calculating the Annual Dose to an Individual Member of the Public.

Response from Applicant: No response is required from the applicant in a license application, but compliance will be examined during inspection. During NRC inspections, licensees must be able to provide documentation demonstrating by measurement or calculation that the total effective dose equivalent to the individual member of the public likely to receive the highest dose from licensed operations is less than 1 mSv (100 mrem) in one year, and any unrestricted area does not exceed 0.02 mSv (2 mrem) in any one hour. See Appendix P for examples of methods to demonstrate compliance.

8.10.7 OPERATING AND EMERGENCY PROCEDURES

Regulations: 10 CFR 20.1406, 10 CFR 20.1906, 10 CFR 20.2201, 10 CFR 20.2202, 10 CFR 20.2203, 10 CFR 21.21(a), 10 CFR 30.50, 10 CFR 39.13, 10 CFR 39.33(b), 10 CFR 39.43, 10 CFR 39.63, 10 CFR 39.67, 10 CFR 39.69, 10 CFR 39.77.

Criteria: The licensee must develop, implement, and maintain operating and emergency procedures or submit a summary of the procedures that addresses the important radiation safety aspects of each procedure to the NRC as part of the application package. Additionally, if well logging and tracer personnel perform specific operations such as leak-testing, semi-annual inspection and maintenance of equipment, and removal and replacement of a sealed source "O" ring, appropriate procedures and instructions for these operations should be included in the applicant's operating and emergency procedures.

Each licensee must develop, implement, and maintain operating and emergency procedures. Operating and emergency procedures' elements must include the items outlined in 10 CFR 39.63. The following is provided as a checklist of important items:

- Instructions for handling and using licensed materials, including sealed sources in wells, without surface casing for protecting fresh water aquifers

- Instructions for maintaining security during storage and transportation

- Instructions to keep licensed material under control and under immediate surveillance during use

- Steps to take to keep radiation exposures ALARA

- Steps to maintain accountability during use

- Steps to control access to work sites

- Steps to take and whom to contact when an emergency occurs

- Instructions for using remote handling tools when handling sealed sources, except low-activity calibration sources and radioactive tracer materials

- Methods and occasions for conducting radiation surveys, including surveys for detecting contamination, as required by 10 CFR 39.67(c) - (e)

- Procedures to minimize personnel exposure during routine use and in the event of an incident, including exposures from inhalation and ingestion of licensed tracer materials

- Methods and occasions for locking and securing stored licensed materials

- Personnel monitoring, including bioassays, and the use of personnel monitoring equipment

- Transportation of licensed materials to field stations or temporary job sites, packaging of licensed materials for transport in vehicles, placarding of vehicles when needed, and physically securing licensed materials in transport vehicles during transportation to prevent accidental loss, tampering, or unauthorized removal

- Procedures for picking up, receiving, and opening packages containing licensed materials, in accordance with 10 CFR 20.1906

- Instructions for the use of tracer materials, including how to decontaminate the environment, equipment, and personnel

- Instructions for maintaining records in accordance with the regulations and the license conditions

- Steps for the use, inspection, and maintenance of sealed sources, source holders, logging tools, injection tools, source handling tools, storage containers, transport containers, and uranium sinker bars, as required by 10 CFR 39.43

- Procedures for identifying and reporting to NRC defects and noncompliance, as required by 10 CFR 21.21(a)

- Actions to be taken if a sealed source is lodged in a well

- Procedures and actions to be taken if a sealed source is ruptured, including actions to prevent the spread of contamination and minimize inhalation and ingestion of licensed materials and actions to obtain suitable radiation survey instruments, as required by 10 CFR 39.33(b)

- Instructions for the proper storage and disposal of radioactive waste

- Procedures for laundering contaminated clothing and for decontaminating equipment and vehicles

- Procedures to be followed in the event of uncontrolled release of radioactive tracer material to the environment, including notification of the RSO, NRC, and other Federal and State Agencies.

Discussion: The purpose of operating and emergency procedures is to provide well logging and tracer personnel, including field flood study personnel, with specific guidance for all operations they will perform. Each topic of importance should be included in the operating and emergency procedures and need not be presented in order. Instructions for non-routine operations, for example, inspection and maintenance of well logging and tracer equipment or conducting calibration of survey instruments, should be included as separate appendices in the application.

Operating and emergency procedures need not specify a particular make and model of survey instrument. Procedures should provide sufficient guidance and instruction for each specific type of well logging or associated equipment. For example, you may submit a single operating procedure for using sealed sources, tracer materials, and isotopes used in field flood operations, provided the unique variances in each operation are addressed in the application.

Operating and emergency procedures or a summary of the procedures that addresses the important radiation safety aspects of each must be submitted to the NRC for review as a part of the application.

Response from Applicant: If applicable to the materials and uses proposed, the licensee must develop, implement, and maintain operating and emergency procedures or submit a summary of the procedures that addresses the important radiation safety aspects of each to the NRC as part of the application package. Applicants should either submit their Operating and Emergency Procedures or an outline or summary as described in 10 CFR 39.13(c) in responding to subsequent sections.

8.10.8 LEAK TESTS

Regulations: 10 CFR 30.53, 10 CFR 39.13(f), 10 CFR 39.35.

Criteria: NRC requires testing of sealed sources containing greater than 3.7 MBq (100 microcuries) of beta/gamma or 0.37 MBq (10 microcuries) of alpha radioactive material in order to determine whether there is any radioactive leakage from sealed sources. Requirements for leak tests are based on the type of radiation (beta/gamma/alpha) escaping from the inner capsule. Records of test results must be maintained.

Discussion: NRC licenses will require the performance of leak tests on sealed sources authorized for well logging at intervals approved by the NRC or an Agreement State and specified in the SSD Registration Sheet. The measurement of the leak test sample is a quantitative analysis requiring that instrumentation used to analyze the sample be capable of detecting 185 becquerels (0.005 microcuries) of radioactivity.

Manufacturers, consultants, and other organizations may be authorized by NRC or an Agreement State either to perform the entire leak test sequence for other licensees or to provide leak test kits to licensees. In the latter case, the licensee is expected to take the leak test sample according to the gauge manufacturer's and the kit supplier's instructions and return it to the kit supplier for evaluation and reporting results. Licensees may also be authorized to conduct the entire leak test sequence themselves.

Response from Applicant:

Do either of the following:

- State: "Leak tests when required by the license will be performed at intervals approved by the NRC or an Agreement State and specified in the Sealed Source and Device Registration Sheet. Leak tests will be performed by an organization authorized by NRC or an Agreement State to provide leak testing services to other licensees or by the licensee using a leak test kit supplied by an organization authorized by NRC or an Agreement State to provide leak test kits to other licensees and according to the kit supplier's instructions."

OR

- State: "Leak testing procedures and analysis will be done by the applicant." Provide the information in supporting a request to perform leak testing. Appendix R may serve as guidance.

OR

- State: "Leak testing will follow the model procedures in Appendix R."

Note: Requests for authorization to perform leak testing and sample analysis will be reviewed on a case-by-case basis and, if approved, NRC staff will authorize via a license condition. Alternative procedures will be evaluated against Appendix R criteria.

References: Draft Regulatory Guide FC 412-4, "Guide for the Preparation of Applications for the Use of Radioactive Materials in Leak-Testing Services," is available from NRC upon request.

8.10.9 MAINTENANCE

Regulations: 10 CFR 39.31, 10 CFR 39.43, 10 CFR 39.49

Criteria: The licensee shall have written procedures for visually inspecting and for maintaining source holders, logging tools, and source handling tools in an operable condition, including labeling. If equipment problems are found, the equipment must be withdrawn from service until repaired. Records of this inspection program are required.

Discussion: Each licensee shall visually check source holders, logging tools, and source handling tools for defects prior to each use to ensure that the equipment is in good working order and that required labeling is present. If defects are found, the equipment must be removed from service until repaired and a record made of the defect and the repairs made prior to returning the equipment for use. At intervals not to exceed 6 months, licensees shall conduct a visual inspection to ensure that no physical damage to equipment is visible and the required labeling is present. Licensees must establish a program for the routine maintenance of source holders, logging tools, inspection tools, source handling tools, storage containers, transport container, injection tools, and uranium sinker bars. If defects are found during the visible inspection or during the routine maintenance, the equipment must be removed from service until repaired and a record made of the defect and any repairs made prior to returning the equipment for use.

Non-routine and special maintenance, e.g., change of O rings on sealed sources or removal of a stuck sealed source, in a manner that could potentially damage or rupture the source, can only be performed by those licensees that have specifically received authorization from the NRC or an Agreement State.

If defects are found as a result of the inspection and maintenance programs, the equipment must be removed from service until repairs are made, and a record of the defect must be retained for 3 years after the defect is found.

Response from Applicant: No response required in the section. Applicants must include in subsequent sections its program for inspection and maintenance of logging equipment and include the program with the Operating and Emergency Procedures.

8.10.9.1 DAILY MAINTENANCE

Regulations: 10 CFR 39.31, 10 CFR 39.43(a), 10 CFR 39.49.

Criteria: The licensee must have written procedures for visually inspecting and maintaining source holders, logging tools, and source handling tools for defects prior to use. This visual inspection is necessary to ensure that the equipment remains in good working condition and is labeled as required.

Discussion: 10 CFR 39.43(a), requires that logging tools, source holders, and source handling tools be checked visually for defects prior to use to ensure that the equipment is in good working condition and is labeled as required. Labeling requirements are specified in 10 CFR 39.31 and 39.49. Instructions in the operating procedures provided to personnel must clearly reflect the regulatory requirement—visual inspections are performed prior to use. Record after the inspection the date, inspector, equipment involved, any defects found, or repairs made. Equipment that fails the inspection and cannot be repaired must be removed from service and returned only after it is successfully repaired.

The licensee must develop, implement, and maintain procedures for visually inspecting and maintaining source holders, logging tools, and source handling tools.

Response from Applicant:

Provide the following:

- Submit a description of procedure(s) for conducting daily visual inspection.

OR

- State that "Visual daily inspections will be conducted and records maintained in accordance with Section 8.10.9.1 of NUREG 1556, Vol. 14, to ensure that well logging equipment is in good working condition and is labeled as required."

8.10.9.2   SEMI-ANNUAL VISUAL INSPECTION AND ROUTINE MAINTENANCE

Regulations: 10 CFR 21.21, 10 CFR 39.31, 10 CFR 39.43(a), 10 CFR 39.43(b), 10 CFR 39.49.

Criteria: Licensees must have written procedures for semiannual visual and routine maintenance of source holders, logging tools, injection tools, source handling tools, storage containers, transport containers, and uranium sinker bars to ensure that the labeling required by 10 CFR Part 39 is legible and that no physical damage to the equipment is visible. Requirements in 10 CFR 21.21 specify, in part, that licensees adopt appropriate procedures to notify NRC of any equipment that is defective or could result in a substantial safety hazard, and additionally, that management be informed as soon as practicable, within 5 working days after the completion of the evaluation.

Discussion: Logging supervisors or assistants are expected to conduct visual inspections and provide routine maintenance activities on source holders, logging tools, injection tools, source handling tools, storage containers, transport containers, and uranium sinker bars to ensure that the labeling required by 10 CFR Part 39.31 for sealed sources and 10 CFR 39.49 for uranium sinker bars is legible, and that no physical damage is visible. If defects are found, the equipment must be removed from service, and a record must be made, listing: the defects, inspection and maintenance operations performed, and the actions taken to correct the defects. As noted in 10 CFR Part 39, instructions for conducting these activities must be included as part of the operating and emergency procedures. Instructions should be tailored to your specific program and to the equipment possessed and used.

Reporting defects to the NRC, in accordance with 10 CFR 21.21, is a management responsibility. The specific mechanism or procedures for reporting to NRC need not be covered in instructions to personnel.

Response from Applicant:

Provide the following:

- Submit a description of procedure(s) for conducting semiannual inspections and routine maintenance of source holders, logging tools, injection tools, source handling tools, storage containers, transport containers, and uranium sinker bars to ensure that the labeling required by 10 CFR Part 39 is legible and that no physical damage is visible.

OR

- State that "Semiannual inspections and routine maintenance will be conducted and records maintained for source holders, logging tools, injection tools, source handling tools, storage containers, transport containers, and uranium sinker bars in accordance with Section 8.10.9.2 of NUREG-1556, Vol. 14, to ensure that well logging equipment is in good working condition with no physical damage evident and that the required labeling is present."

8.10.9.3 MAINTENANCE REQUIRING SPECIAL AUTHORIZATION

Regulations: 10 CFR 39.43(c), 10 CFR 39.43(d), 10 CFR 39.43(e).

Criteria: Certain maintenance procedures on sealed sources or holders that contain sealed sources are prohibited, unless a written procedure has been approved and the licensee is specifically authorized by the NRC or an Agreement State to perform these operations.

Discussion: Activities that are prohibited, unless a written procedure has been reviewed and approved by NRC or an Agreement State, include:

- Removing a sealed source from a source holder or logging tool

- Preventive maintenance activities on sealed sources or holders that may be necessary when using certain types of logging tools, including removing and replacing O-rings (see Figure 8.12 below)

- Removing a sealed source that is stuck in a source holder or logging tool, e.g., any situation where tools are required to remove the stuck source.

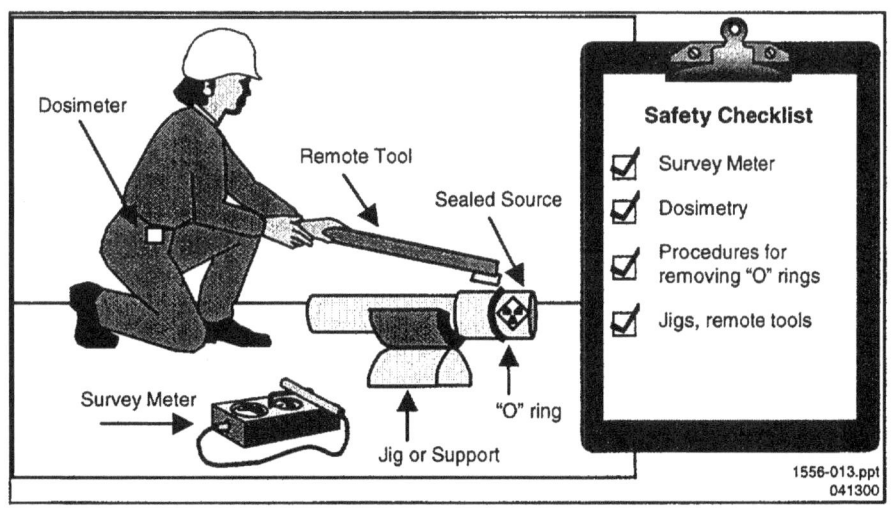

Figure 8.12 Maintenance, Cleaning, and O-Ring Replacement.

Response from Applicant:

- Statement that "Prohibited activities described in Section 8.10.9.3 of NUREG-1556, Vol. 14 will not be conducted unless approved by the NRC."

OR

- Submit detailed procedures for any prohibited activities, including radiation safety precautions that individuals will be expected to follow when performing these tasks and the minimum qualifications of these individuals. Each different task must be described. Should a procedure require the removal of the sealed source from the holder before performing any maintenance on the holder, applicants should describe the removal procedures.

Note: Equipment manufacturers can provide information concerning maintenance and source removal procedures. In some cases, certain maintenance operations should only be performed by the manufacturer or individuals who are licensed by NRC or an Agreement State to provide these services.

8.10.10 TRANSPORTATION

Regulations: 10 CFR 20.1101, 10 CFR 30.41, 10 CFR 30.51, 10 CFR 39.31, 10 CFR 71.5, 10 CFR 71.12, 10 CFR 71.13, 10 CFR 71.14, 10 CFR 71.37, 10 CFR 71.38, 10 CFR 71.47, Subpart H of 10 CFR Part 71, 49 CFR Parts 171-178.

Criteria: Applicants must develop, implement, and maintain safety programs for transport of radioactive material to ensure compliance with NRC and Department of Transportation (DOT) regulations.

Discussion: Licensees should consider the safety of all individuals who may either handle or come into contact with transport containers or packages containing licensed material. The primary consideration in packaging licensed material should be to ensure that the package integrity is not compromised during transport and that the radiation levels or removable contamination levels at the package surfaces meet the regulatory requirements of 10 CFR 71.47. In all cases, ALARA concerns are addressed prior to, during, and after transporting any radioactive material.

Note: Licensees shipping radioactive waste for disposal must prepare appropriate documentation as specified in 10 CFR Part 20 and Appendix S.

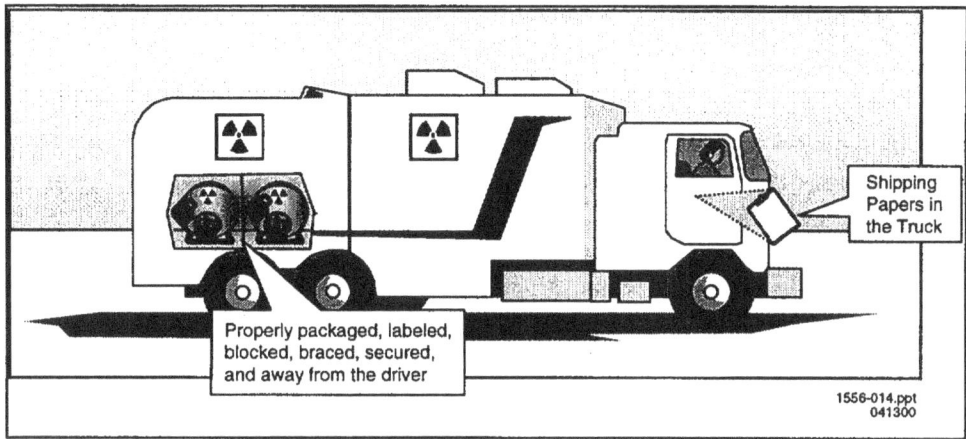

Figure 8.13 Transportation. *Licensees often transport their equipment and radioactive materials, including sealed sources and tracer materials, to and from job sites and must ensure compliance with DOT regulations.*

Discussion: Figure 8.13 illustrates some DOT requirements often overlooked by well logging, tracer, and field flood study licensees. During an inspection, NRC uses the provisions of 10 CFR 71.5 and a Memorandum of Understanding with DOT to examine and enforce transportation requirements applicable to well logging, tracer and field flood study licensees. Appendix S lists major DOT regulations and provides a sample shipping paper.

Figure 8.14 Transport Container.

Response from Applicant: No response is needed from applicants during the licensing phase. Transportation issues are reviewed during inspections.

References: "A Review of Department of Transportation Regulations for Transportation of Radioactive Materials (1983 revision)" can be obtained be calling DOT's Office of Hazardous Material Initiatives and Training at (202) 366-4425.

8.10.11 MINIMIZATION OF CONTAMINATION

Regulations: 10 CFR 20.1406, 10 CFR 39.33(a), 10 CFR 39.35(d), 10 CFR 39.67(c)-(e), 10 CFR 39.69.

Criteria: Applicants for new licenses must describe how facility design and procedures for operation will minimize, to the extent practicable, contamination of the facility and the environment, facilitate eventual decommissioning, and minimize, to the extent practicable, the generation of radioactive waste.

Discussion: When designing facilities and developing procedures for their safe use, applicants should plan ahead and consider how to minimize radioactive contamination during operation, decontamination and decommissioning efforts, and radioactive waste generation. When submitting new applications, applicants should consider the following:

- Implementation of and adherence to good health physics practices while performing operations

- Minimization of distance to areas, to the extent practicable, where licensed materials are used and stored

- Maximization of survey frequency, within reason, to enhance detection of contamination

- Segregation of radioactive material in waste storage areas

- Segregation of sealed sources and tracer materials to prevent cross-contamination

- Separation of radioactive material from explosives

- Separation of potentially contaminated areas from clean areas by barriers or other controls.

Sealed sources found to be leaking in excess of 185 bequerels (.005 microcuries) of removal contamination must be immediately withdrawn from use and placed in a safe storage location until disposed of according to NRC requirements. Special authorization must be granted by NRC to applicants to decontaminate a facility contaminated by a leaking sealed source. Approval granted in a license by NRC or an Agreement State to provide these specialized services minimizes the spread of contamination and reduces radioactive waste associated with decontamination efforts.

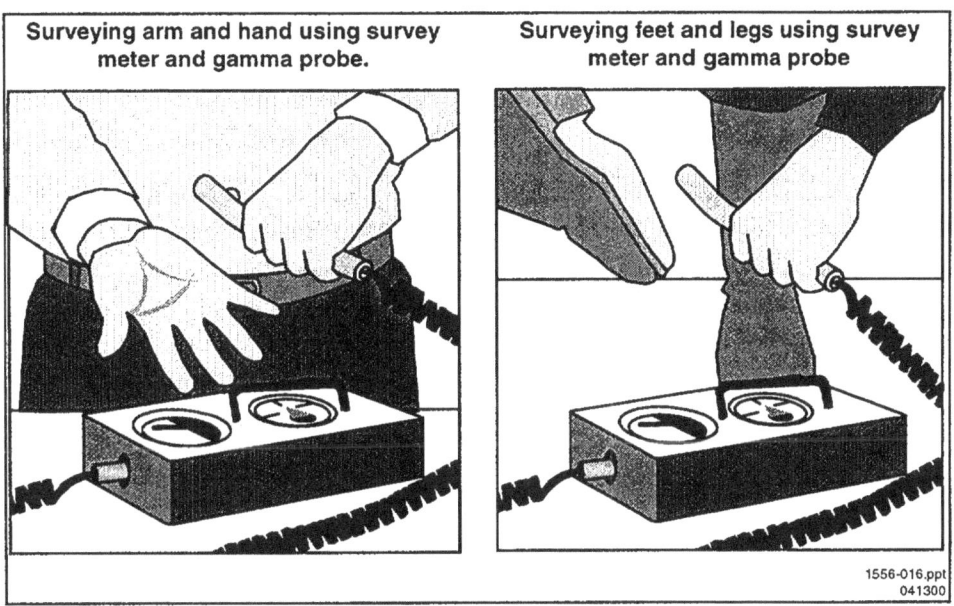

Figure 8.15 Personnel Surveys.

Response from Applicant:

• The applicant does not need to provide a response to this item under the following conditions and NRC will consider that the above criteria have been met if the applicant's responses meet the criteria in the following sections: "Facilities and Equipment," "Radiation Safety Program - Tracer Studies," "Radiation Safety Program - Operating and Emergency Procedures," and "Radiation Safety Program - Waste Management."

AND

• Decontamination of the facility and/or the sealed source requires special authorization from the NRC or an Agreement State.

OR

• The licensee should submit its procedures to perform major decontamination activities if it intends to perform the activity rather than contracting the work to a licensed entity.

8.10.12 SEALED SOURCES

8.10.12.1 DRILL-TO-STOP LARGE SEALED SOURCES

Regulations: 10 CFR 39.13, 10 CFR 39.63.

Criteria: Licensee must develop and follow instructions to be used by logging personnel when using licensed sealed radioactive sources in drill-to-stop well logging operations.

Unlike measurement while drilling (MWD) or logging while drilling (LWD) operations where well logging operations occur concurrent with the drilling operations, drill-to-stop (DTS) well logging operations require that all drilling operations cease and that parts of the drilling apparatus, including all of the drill stem, be removed to provide access to the well bore. The well logging tool containing one or more sealed sources is then lowered into the well bore to obtain information about the well or adjacent oil, gas, mineral, groundwater, or geological formations.

Figure 8.16 Drill-to-stop Well Logging Operations.

Discussion: Operating and Emergency procedures that cover the use of sealed sources in DTS well logging operations must be developed and implemented.

Applicants who request authorization to use sealed sources in DTS well logging operations in well bores without a surface casing should describe the procedures to be followed necessary to ensure that a sealed source does not become lodged in the well bore. Examples of acceptable procedures include:

- Obtaining specific knowledge of the borehole conditions from the drilling team or company

- First running a caliper log to show the hole is open or to find problem areas

- First running a tool without a radioactive source to show it can be freely removed

- Placing a temporary casing in sections of the hole giving problems.

Instructions in drill-to-stop well logging activities should include procedures for using appropriate remote handling tools for handling sealed sources. If only certain handling tools are to be used with particular sealed sources, instructions should clearly address which handling tool is required for each specific sealed source.

Response from Applicant:

- Submit operating and emergency procedures for conducting DTS well logging operations

OR

- Submit an outline or summary that addresses important radiation safety aspects of its Operating and Emergency Procedures when conducting DTS well logging operations.

8.10.12.2 MEASUREMENT WHILE DRILLING, LOGGING WHILE DRILLING

Regulations: 10 CFR 39.13, 10 CFR 39.63.

Criteria: Licensees must develop and follow procedures to be used by logging personnel when using licensed sealed radioactive sources in MWD or LWD well logging operations.

MWD or LWD well logging operations occur during the drilling of the well bore and do not require that the drill stem or other equipment be removed from the well. MWD or LWD requires that the well logging tool containing one or more sealed sources be located above the drilling stem to obtain information about the well or adjacent oil, gas, mineral, groundwater, or geological formations while the well drilling operation continues uninterrupted. Both MWD and LWD activities can be conducted at the same time drilling operations are occurring. Downhole recorded data from MWD or LWD sensors is transmitted to the surface through the use of mud telemetry.

Discussion: Operating and Emergency procedures that cover the use of sealed sources in MWD or LWD well logging operations must be developed and implemented.

Instructions in MWD and LWD well logging activities should include procedures for using appropriate remote handling tools for handling sealed sources. If only certain handling tools are to be used with particular sealed sources, instructions should clearly address which handling tool is required for each specific sealed source.

Response from Applicant:

- Submit operating and emergency procedures for conducting MWD and LWD well logging activities

OR

- Submit an outline or summary that addresses important radiation safety aspects of Operating and Emergency Procedures when conducting MWD and LWD well logging activities.

8.10.12.03 ENERGY COMPENSATION SOURCES

Regulations: 10 CFR 39.13, 10 CFR 39.35, 10 CFR 39.37, 10 CFR 39.39, 10 CFR 39.41, 10 CFR 39.51, 10 CFR 39.63.

Criteria: Energy compensation sources (ECSs) used in well logging operations are low-activity special form singly or doubly encapsulated sources containing less than or equal to 3.7 MBq (100 microcuries) of byproduct material. ECSs are used as reference or calibration standards for stabilizing and calibrating conventional, LWD, or MWD well logging tools.

Discussion: ECSs are not considered well logging sealed sources and are not required to satisfy the requirement for well logging sealed sources. As a result, ECSs are:

- Exempt, in most instances, from leak testing requirements, per 10 CFR 39.35(e). ECSs requiring leak testing must be tested at intervals not to exceed 3 years.

- Exempt from abandonment requirements when only ECSs less than or equal to 3.7 MBq (100 microcuries) remain in the abandoned tool.

- Exempt from the performance requirements of sealed sources used in well logging operations.

- Exempt from the monitoring requirements during source recovery operations when only ECSs less than or equal to 3.7 MBq (100 microcuries) remain in a well logging tool that is lodged in a well.

- Exempt from all requirements in 10 CFR Part 39, with the exceptions of physical inventory and records of use. Requirements established in other parts of NRC regulations (e.g., 10 CFR Part 20, 10 CFR Part 30) are still applicable to possession and use of byproduct material contained in ECSs.

- If a surface casing is not used to protect fresh water aquifers, see 10 CFR 39.53 for applicable requirements.

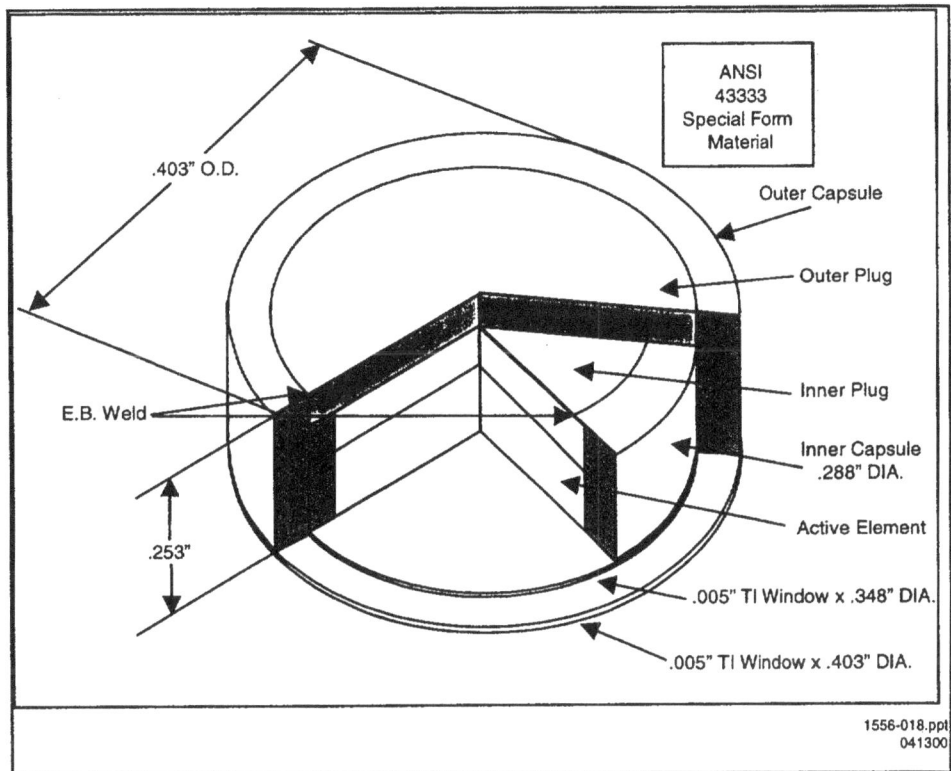

Figure 8.17 Singly Encapsulated ECS Sealed Source.

Response from Applicant:

- Submit Operating and Emergency Procedures for using and handling ECSs

OR

- Submit an outline or summary that addresses important radiation safety aspects of operating and emergency procedures when using or handling ECSs. The summary must include:

    — Instructions for testing ECSs requiring leak tests at intervals not to exceed 3 years

    — Instructions for conducting physical inventories of ECSs at least every 6 months

    — A record system for maintaining inventory records required by 10 CFR 39.37

    — A record system for maintaining records of use for ECSs.

OR

- Submit alternative procedures for NRC's review.

8.10.12.4   USE OF SEALED SOURCES OR NEUTRON GENERATORS IN FRESH WATER AQUIFERS

Regulations: 10 CFR 39.63.

Criteria: The licensee is prohibited from using sealed sources or neutron generators in fresh water aquifers unless the licensee requests and receives written permission from the NRC.

Discussion: Use of radioactive materials in fresh water aquifers is a prohibited activity. Authorizing to use sealed sources or neutron generators in fresh water aquifers requires that OE procedures include the following information:

- Obtaining specific knowledge of the borehole conditions from the drilling team or company

- First running a caliper log to show the hole is open or to find problem areas

- First running a tool without a radioactive source to show it can be freely removed

- Placing a temporary casing in sections of the hole giving problems.

Response from Applicant: No response is required from the licensee unless it requests authorization for the prohibited activity.

8.10.13 TRACER STUDIES

8.10.13.1   TRACER STUDIES IN SINGLE WELL APPLICATIONS

Regulations: 10 CFR 39.45, 10 CFR 39.63.

Criteria: Applicants must develop, implement, and maintain safety programs for the use of unsealed material for tracer studies in single wells.

Discussion: Applicants' operating and emergency procedures should address the following concerns:

- Methods and occasions for conducting radiation surveys

- Methods and occasions for locking and securing tracer materials

- Personnel monitoring and the use of personnel monitoring equipment

- Transportation to temporary job sites and field stations, including the packaging and placing of tracer materials in vehicles, placarding of vehicles, and securing of tracer materials during transportation

- Procedures for minimizing exposure to members of the public and occupationally exposed individuals in the event of an accident

- Maintenance of records at field stations and temporary job sites

- Use, inspection, and maintenance of equipment (injector tools, remote handling tools, transportation containers, etc.)

- Procedures to be used for picking up, receiving, and opening packages containing radioactive material

- Decontamination of the environment, equipment, and personnel

- Notifications of proper personnel in the event of an accident.

Response from Applicant: No response is required to this section, provided that the elements listed above are contained in other sections.

8.10.13.2 FIELD FLOOD AND SECONDARY RECOVERY APPLICATIONS (TRACER STUDIES IN MULTIPLE WELLS)

Regulations: 10 CFR 39.45, 10 CFR 39.63, 10 CFR 51.21, 10 CFR 51.22, 10 CFR 51.30, 10 CFR 51.60, 10 CFR 51.66.

Criteria: Applicants must develop, implement, and maintain safety programs for the use of unsealed material for tracer studies in multiple wells (field flood studies). Refer to Appendix F in developing step-by-step instructions for tracer personnel in performing field flood tracer studies for multiple wells.

Field flood study activities where licensed material is intentionally released into the environment require an environmental assessment (EA) in accordance with the provisions of 10 CFR 51.21.

NRC has determined that a full environmental assessment is not required, provided the amount of material requested for each isotope is within the generic bounding criteria established in Table 2.1 of NUREG/CR-3467, "Environmental Assessment of the Use of Radionuclides as Tracers in the Enhanced Recovery of Oil and Gas," dated November 1983. For copies of NUREG/CR-3467, see the Notice of Availability on the inside front cover of this report.

Discussion: Applicants should address the following when requesting field flood and secondary recovery applications:

- Agreement with well operator or owner

- Field flood study project design

- Pre-injection phase of the field flood project

- Injection phase

- Post-injection phase

- Emergency procedures

- Reporting and record keeping requirements

- Waste management

- Methods and occasions for conducting radiation surveys

- Methods and occasions for locking and securing tracer materials

- Personnel monitoring and the use of personnel monitoring equipment

- Transportation to temporary job sites and field stations, including the packaging and placing of tracer materials in vehicles, placarding of vehicles, and securing tracer materials during transportation

- Procedures for minimizing exposure to members of the public and occupationally exposed individuals in the event of an accident

- Maintenance of records at field stations and temporary job sites

- Use, inspection and maintenance of equipment (injector tools, remote handling tools, transportation containers, etc.)

- Procedures to be used for picking up, receiving, and opening packages containing radioactive material

- Decontamination of the environment, equipment, and personnel

- Notifications of proper personnel in the event of an accident.

Response from Applicant:

- Statement that "Field flood studies using tracer materials will not be conducted unless authorized specifically by license conditions."

OR

- Licensees requesting authorization to conduct field flood studies in the enhanced recovery of oil and gas wells, should provide the information in Appendix F.

8.10.13.3 TRACER STUDIES IN FRESH WATER AQUIFERS

Regulations: 10 CFR 39.45, 10 CFR 51.22, 10 CFR 51.30,10 CFR 51.60, 10 CFR 51.66.

Criteria: Applicants must develop, implement, and maintain a safety program for using tracer materials in fresh water aquifers. Licensees may not knowingly inject licensed material into a freshwater aquifer unless specifically authorized to do so by the Commission.

Discussion: In 10 CFR Part 51.22, NRC specifies the criteria for categorical exclusions. When one or more of the criteria for a categorical exclusion are satisfied, the applicant or licensee is relived from the requirements for preparing an environmental impact statement. This then relieves the NRC from the requirement of preparing an environmental assessment prior to the issuance, amendment, or renewal of licenses authorizing the use of radioactive tracers in well logging procedures authorized under 10 CFR Part 39. However, the intentional release of licensed radioactive material directly to the environment as a result of a research or other study is not categorically excluded. The Commission specifies in 10 CFR 51.21 and 51.22(b) that in special circumstances and on its own initiative or on the request of any interested individual or party, an environmental assessment on an action normally covered by a categorical exclusion could be required.

NRC, in accordance with 10 CFR 39.45(b), prohibits the intentional injection of licensed tracer material into a fresh water aquifer unless the individual is specifically authorized by the Commission to perform this activity. NRC staff position concerning the intentional injection of licensed tracer material authorized under 10 CFR Part 39 into a fresh water aquifer requires the preparation of an environmental report by the licensee or applicant. Well logging applicants and applicants requesting field flood studies should refer to 10 CFR Part 51.45 and prepare an environmental report. Authorizing an applicant to conduct tracer studies in accordance with 10 CFR 39 in fresh water aquifers would require NRC's assessment of an environmental report and a "finding of no significant impact" by the NRC staff.

Authorizing field flood studies that require the applicant to intentional inject licensed tracer material into a fresh water aquifer would require that an environmental report be prepared by the applicant and an environmental assessment be made by the NRC. Field flood study applicants are charged at full cost fee based on the professional staff time expended as described in footnote e.3. to 10 CFR 170.31. Individuals planning activities of this nature should contact NRC well in advance of scheduled use.

Note: NRC's completion of an environmental assessment, based on the level of complexity, can require several months to review, approve, and publish in the *Federal Register* for comments.

Response from Applicant:

- State that: "We will not knowingly inject tracer material into a fresh water aquifer."

OR

- Applicants requesting authorization to inject licensed radioactive material into a fresh water aquifer must provide their reasons for performing the study and procedures to protect their occupationally exposed workers and the public. For tracer and field flood studies, licensees must also provide an environmental report containing the information outlined in 10 CFR 51.45. Applications require that NRC conduct an assessment and prepare an environmental impact statement. Authorization to inject licensed radioactive material into a fresh water aquifer requires that applicants provide procedures to safeguard the public, licensee personnel, and the environment.

8.10.14 RADIOACTIVE COLLAR AND SUBSIDENCE OR DEPTH CONTROL MARKERS

Regulations: 10 CFR 30.71, 10 CFR 39.47, 10 CFR 39.37.

Criteria: Radioactive markers usually used as pipe collar markers include wires, tape, nails, etc. Applicants can use radioactive markers only where each individual marker contains quantities of licensed material not exceeding the quantities identified in 10 CFR 30.71, Schedule B. Radioactive markers must be physically inventoried at intervals not to exceed 6 months, as specified in 10 CFR 39.37.

Discussion: Operating and emergency procedures must include a commitment that radioactive markers can be used only where each individual marker contains quantities of licensed material not exceeding the quantities identified in 10 CFR 30.71, Schedule B. However, licensees are not restricted to using only one marker, and may use multiple markers in each pipe joint, provided each individual marker (wires, tape, nails, etc.) is not greater than the quantities identified in

10 CFR 30.71. Additionally, provisions must be included in the operating and emergency procedures to ensure that radioactive markers undergo physical inventories at intervals not to exceed 6 months, as specified in 10 CFR 39.37.

Note: Subsidence or depth control markers that use quantities greater that those authorized by 10 CFR 39.47 must be approved or registered by the NRC or an Agreement State in an SSD Registration Certificate.

Response from Applicant:

• State that: "We will only use radioactive markers where each individual marker contains only quantities of licensed material not exceeding the quantities identified in 10 CFR 30.71, Schedule B, as described in Section 8.10.14 of NUREG-1556, Vol.14."

OR

• Submit procedures for using radioactive markers that are in excess of the quantities in Section 8.10.14 of NUREG-1556, Vol.14.

8.10.15 NEUTRON ACCELERATORS USING LICENSED MATERIAL

Regulations: 10 CFR 20.1301, 10 CFR 20.1302, 10 CFR 20.1601, 10 CFR 20.1602, 10 CFR 39.55.

Criteria: Applicants authorized to use a neutron generator (particle accelerator) containing a tritium source, should include operating and emergency procedures for the proper handling and use of the accelerator targets or tubes containing radioactive materials. Because the neutron radiation produced from particle accelerators containing byproduct materials is categorized as machine-produced radiation, it is subject to individual State, not NRC, regulatory authority. Nonfederal applicants using neutron generators should contact the appropriate State for additional information.

> *Note:* Machine-produced radiation dose is additive to the dose from NRC-regulated materials when assessing total occupational dose occurring during a specified time interval.

Discussion: Neutron generators (accelerators) are used in the well logging industry as a source of neutrons. Most accelerators use tritium gas sealed in a glass tube or plated on a target or disc. Neutron generator target sources, in most instances, contain less than 110 GBq (30 curies) of tritium.

Neutron generator tubes are not considered well logging sealed sources and are not required to satisfy the requirement for well logging sealed sources. As a result, neutron generator tubes containing less than 110 GBq (30 curies) of tritium are:

• Exempt from abandonment requirements

• Exempt from leak test requirements

• Exempt from the performance requirements of sealed sources used in well logging operations

• Not exempt if a tritium neutron generator for target source is greater than 100 GBq (30 curies) or is used in a well without a surface casing to protect fresh water aquifers.

Section Guidance: Applicants using a neutron generator (particle accelerator) should include handling procedures that address contamination. Operating and Emergency procedures should instruct individuals in the handling of contamination resulting from the routine use, initial installation, replacement, or accidental damage of the targets or glass tubes. Refer to 10 CFR 39.55 for applicable requirements for using neutron generators.

Response from Applicant:

- State that: "We will not use neutron generators (accelerators) in our well logging operations."

OR

- State that "We will use neutron generators (accelerators) in accordance with the criteria in Section 8.10.15 of NUREG-1556, Vol. 14."

8.10.16 DEPLETED URANIUM SINKER BARS

Regulations: 10 CFR 39.43(b), 10 CFR 39.49, 10 CFR 39.67, 10 CFR 40.25, 10 CFR 40.51.

Criteria: Depleted uranium sinker bars are both generally licensed and specifically licensed. Most well logging licensees acquire depleted uranium sinker bars under the provisions of 10 CFR 40.25 and then file Form NRC 244, "Registration Certificate — Use of Depleted Uranium Under General License." Specifically licensed material must be physically inventoried and visually inspected for labeling and physical damage.

Discussion:

Depleted Uranium Sinker Bars Authorized Under *General License*:

Certain devices are authorized by NRC for distribution to persons who are generally licensed for the use of certain industrial products or devices containing depleted uranium for the purpose of providing a concentrated mass in a small volume. Uranium sinker bar devices can be acquired by the users under the provisions of 10 CFR 40.25 without obtaining a specific license from NRC; however, when acquired under the provisions of a general license, individuals must file Form NRC 244, "Registration Certificate — Use of Depleted Uranium Under General License." Generally licensed sinker bars are exempt from 10 CFR Parts 19, 20, and 21. Regulatory requirements that apply to such devices possessed under a general license are stated in 10 CFR 40.25. While operating under the provision of a general license for these types of devices, general licensees must:

- Not introduce uranium sinker bars into a chemical, physical, or metallurgical treatment or process, except as a treatment for restoration of any plating or covering

- Not abandon uranium sinker bars

- Transfer only to individuals authorized under the provisions of 10 CFR 40.51

- Notify NRC within 30 days of the transfer of depleted uranium sinker bars.

Depleted Uranium Sinker Bars Authorized under a *Specific License*:

While operating under the provision of a specific license for these types of devices, specific licensees must:

- Physically inventory the uranium sinker bars at intervals not to exceed 6 months

- Visually inspect before use for proper labeling, "CAUTION - RADIOACTIVE DEPLETED URANIUM" and "NOTIFY CIVIL AUTHORITIES (or COMPANY NAME) IF FOUND," and at intervals not to exceed 6 months

- Visually inspect for physical damage and conduct routine maintenance at intervals not to exceed 6 months, as specified in 10 CFR 39.43(b)

- Remove bars from use if found defective, until repaired or disposed

- Record information specified in 10 CFR 39.43(b).

Response from Applicant:

- State that: "Depleted uranium sinker bars will be obtained under the provisions of a general license per 10 CFR 40.51, and registration form NRC Form 244 will be filed, as required."

OR

- State that: "Depleted uranium sinker bars will not be obtained under the provision of a general license per 10 CFR 40.51 (general license)."

AND

- State that: "Uranium sinker bars will be possessed and inspected as specified in Section 8.10.16 of NUREG-1556, Vol. 14."

AND

- Specify the number of kilograms of materials requested.

8.11   ITEM 11: WASTE MANAGEMENT

Regulations: 10 CFR 20.1904, 10 CFR 20.2001, 10 CFR 20.2002, 10 CFR 20.2003, 10 CFR 20.2004, 10 CFR 20.2005, 10 CFR 20.2006, 10 CFR 20.2007, 10 CFR 20.2108, 10 CFR 30.41, 10 CFR 30.51.

Criteria: Radioactive waste must be disposed of in accordance with regulatory requirements and license conditions and/or transferred to an authorized recipient. Authorized recipients are the original manufacturer, distributor, a commercial firm licensed by NRC or an Agreement State to accept radioactive waste from other persons, or in the case of sealed sources, transferred to another specific licensee authorized to possess the licensed material (i.e., a transferees' license specifically authorizes the same radionuclide, chemical or physical form, and in most instances, the same use). Records of transfer and waste disposal must be maintained.

Before transferring any radioactive material, including radioactive waste, a licensee must verify that the recipient is properly authorized to receive the specific type of material using one of the methods described in 10 CFR 30.41. In addition, all packages containing radioactive waste must be prepared and shipped in accordance with NRC and DOT regulations. Records of transfer and disposal must be maintained as required by 10 CFR 30.51.

Discussion: Radioactive waste generated when conducting licensed activities may include: sealed sources, used or unused radioactive tracer materials, and unusable items contaminated with radioactive tracer materials (e.g., absorbent paper, gloves, bottles, etc.).

Unsealed radioactive waste must be stored in strong, tight containers (e.g., thick plastic bags, boxes, barrels, etc.) to prevent the spread of contamination, and sealed sources should be stored in their corresponding transport containers or in a downhole storage bunker until their disposal. The integrity of the radioactive waste containers must be assured, and the containers, while in storage, must have the appropriate warning label specified in 10 CFR Part 20. Radioactive waste must be secured against unauthorized access or removal. Depending on the radioactive half-life of the material, NRC requires disposal of well logging sealed sources and tracer materials generated at licensees' facilities by one or more of the following methods:

Tracer Material with a Half-Life of 120 Days or Less:

- Decay-in-storage (DIS)
- Transfer to an authorized recipient
- Release into sanitary sewerage
- Obtaining prior approval of NRC of any alternate method
- Release in effluents to unrestricted areas, other than into sanitary sewerage
- Incineration.

Tracer Material with a Half-Life Greater Than 120 Days:

- Transfer to an authorized recipient
- Release into sanitary sewerage
- Extended interim storage
- Obtaining prior approval of NRC of any alternate method

- Release in effluents to unrestricted areas, other than into sanitary sewerage

- Incineration.

Sealed Sources with a Half-Life of 120 Days or Less:

- Transfer to an authorized recipient

- DIS

- Extended interim storage.

Sealed Sources with a Half-Life Greater Than 120 Days:

- Transfer to an authorized recipient.

Licensees may choose any one or more of these methods to dispose of their radioactive waste. NRC's experience indicates that most well logging tracers are stored or disposed of by a combination of methods, transfer to an authorized recipient and decay-in-storage being the most frequently used. Applicants requesting authorization to dispose of radioactive tracer waste by incineration should first refer to Policy and Guidance Directive PG 8-10, "Disposal of Incinerator Ash as Ordinary Waste," dated January 1997, and contact the appropriate Regional Office of the NRC for guidance. Applicants should note that compliance with NRC regulations does not relieve them of their responsibility to comply with any other applicable Federal, State, or local regulations. Some types of radioactive waste used in tracer operations and in "labeled frac sands" may include additional chemical hazards. This type of waste is designated as "mixed waste" and requires special handling and disposal.

Applicants should describe in detail their program for management and disposal of radioactive waste, including mixed waste, if applicable. A waste management program should include procedures for handling waste; specify the requirements for safe and secure storage; and describe how to characterize, minimize, and dispose of all types of radioactive waste, including, where applicable, mixed waste. Appropriate training should be provided to waste handlers. Regulation 10 CFR 30.51 requires, in part, that licensees maintain all appropriate records of disposal of radioactive waste. The U.S. Environmental Protection Agency (EPA) issued guidance for developing a comprehensive program to reduce hazardous waste that, in many instances, may also include radioactive waste as a contaminant. NRC transmitted these guidelines to licensees in IN-94-23, "Guidance to Hazardous, Radioactive, and Mixed Waste Minimization Program," dated March 1994.

Disposal By Decay-in-Storage (DIS)

NRC has concluded that materials with half-lives of less than or equal to 120 days are appropriate for DIS. The minimum holding period for decay is ten half-lives of the longest-lived radioisotope in the waste with a half-life of 120 days or less. Such waste may be disposed of as ordinary trash if radiation surveys (performed in a low background area and without any interposed shielding) of the waste at the end of the holding period indicate that radiation levels are indistinguishable from background. All radiation labels must be defaced or removed from containers and packages prior to disposal as ordinary trash. If the decayed waste is compacted, all labels that are visible in the compacted mass must also be defaced or removed.

Applicants should assure that adequate space and facilities are available for the storage of such waste. Licensees can minimize the need for storage space, if the waste is segregated according to physical half-life. Waste containing radioisotopes with physical half-lives 120 days or less may be segregated and stored in a container and allowed to decay for at least ten half-lives based on the longest-lived radioisotope in the container. Waste management procedures should include: (a) methods of segregating waste by physical half-lives of 120 days or less, greater than 120 days; methods of surveying waste prior to disposal to confirm that waste above background levels is not inadvertently released; and maintenance of records of disposal. Disposal records for DIS should include the date when the waste was put in storage for decay, date when ten half-lives of the longest-lived radioisotope had transpired, date of disposal, and results of final survey taken prior to disposal to ordinary trash. Additionally, a model procedure for disposal of radioactive waste by DIS, which incorporates the above guidelines, is provided in Appendix T.

Release Into Sanitary Sewerage

10 CFR 20.2003 authorizes disposal of radioactive waste by release into sanitary sewerage if each of the following conditions is met:

- Material is readily soluble (or is easily dispersible biological material) in water

- Quantity of licensed material that the licensee releases into the sewer each month averaged over the monthly volume of water released into the sewer does not exceed the concentration specified in 10 CFR Part 20, Appendix B, Table 3

- If more than one radioisotope is released, the sum of the ratios of the average monthly discharge of a radioisotope to the corresponding limit in 10 CFR Part 20, Appendix B

- Table 3 cannot exceed unity

- Total quantity of licensed material released into the sanitary sewerage system in a year does not exceed 185 GBq (5 Ci) of H-3, 37 GBq (1 Ci) of C-14, and 37 GBq (1 Ci) of all other radioisotopes combined.

Licensees are responsible to demonstrate that licensed materials discharged into the sewerage system are indeed readily dispersible in water. NRC IN 94-07, "Solubility Criteria for Liquid Effluent Releases to Sanitary Sewerage Under the Revised 10 CFR 20," dated January 1994, provides the criteria for evaluating solubility of liquid waste. Careful consideration should be given to the possibility of reconcentration of radioisotopes that are released into the sewer. NRC alerted licensees to the potentially significant problem of reconcentration of radionuclides released to sanitary sewerage systems in IN 84-94, "Reconcentration of Radionuclides Involving

Discharges into Sanitary Sewerage Systems Permitted Under 10 CFR 20.203 (now 10 CFR 20.2003)," dated December 1984.

Applicants electing to use this type of disposal should provide procedures that will ensure that all releases of radioactive waste into the sanitary sewerage meet the criteria stated in 10 CFR 20.2003 and do not exceed the monthly and annual limits specified in regulations. Licensees are required to maintain accurate records of all releases of licensed material into the sanitary sewerage. A model program for disposal of radioactive waste via sanitary sewer is described in Appendix T.

> *Note:* 10 CFR Part 20 prohibits the disposal of radioactive materials via a sewage treatment facility, septic system or leach field owned or operated by the licensee.

Transfer to an Authorized Recipient

Licensees may transfer radioactive waste to an authorized recipient for disposal. However, it is the licensee's responsibility to verify that the intended recipient is authorized to receive the radioactive waste prior to making any shipment. Waste generated at well logging and tracer facilities generally consists of low specific activity (LSA) material. The waste must be packaged in DOT-approved containers for shipment, and each container must identify the radioisotopes and the amounts contained in the waste. Additionally, packages must comply with the requirements of the particular burial site's license and State requirements. Each shipment must comply with all applicable NRC and DOT requirements. In some cases, the waste handling contractor may provide additional guidance and requirements to licensees for packaging and transportation; however, the licensee is ultimately responsible for ensuring compliance with all applicable regulatory requirements.

The shipper must provide all information required in NRC's Uniform Low-Level Radioactive Waste Manifest and transfer this recorded manifest information to the intended recipient. Each shipment manifest must include a certification by the waste generator. Each person involved in the transfer for disposal and disposal of waste, including waste generator, waste collector, waste processor, and disposal facility operator, must comply with NRC's Uniform Low-Level Radioactive Waste Manifest.

Licensees should implement procedures to reduce the volume of radioactive waste for final disposal in an authorized low-level radioactive waste (LLW) disposal facility. These procedures include volume reduction by segregating, consolidating, compacting, or allowing certain waste to decay in storage. Waste compaction or other treatments can reduce the volume of radioactive waste, but such processes may pose additional radiological hazards (e.g., airborne radioactivity) to workers and members of the public. The program should include adequate safety procedures to protect workers, members of the public, and the environment.

Applicants may request alternate methods for the disposal of radioactive waste generated at their facilities. Such requests will be handled on a case-by-case basis and require that the applicant provide additional site-specific information. In most instances, requests for alternate methods of disposal must describe the types and quantities of waste containing licensed material, physical and chemical properties of the waste that may be important to making a radiological risk assessment, and the proposed manner and conditions of waste disposal. Additionally, the applicant must submit its analysis and evaluation of pertinent information specific to the affected environment, including the nature and location of other affected facilities, and provide an outline of its procedures to ensure that radiation doses are maintained ALARA and within regulatory limits.

Because of the difficulties and costs associated with disposal of sealed sources, e.g., sealed sources containing americium-241, applicants should preplan disposal. Applicants may want to consider contractual arrangements with the source supplier as part of a purchase agreement.

Extended Interim Storage

Prior to requesting extended interim storage of radioactive waste materials, and this only as a last resort, licensees should exhaust all possible alternatives for disposal of radioactive waste. The protection of occupationally exposed workers or the public is enhanced by disposing of radioactive waste, rather than storing it. In addition, licensees may find it more economical to dispose of radioactive waste than to store it on-site. As available burial ground capacity decreases, cost of disposal of radioactive waste most likely will continue to increase. Other than DIS, LLW should be stored only when disposal capacity is unavailable and for no longer than is necessary. NRC IN 90-09, "Extended Interim Storage of Low-Level Radioactive Waste by Fuel Cycle and Materials Licensees," dated February 1990 and NRC IN 93-50, "Extended Storage of Sealed Sources," dated July 1993, provides guidance to licensees for requesting an amendment to authorize extended interim storage of both sealed and unsealed LLW.

Response from Applicant:

A statement that:

1. "We will use the model waste procedures published in Appendix T to NUREG-1556, Vol. 14, 'Program-Specific Guidance About Well Logging, Tracer, and Field Flood Study Licenses', dated April 2000."

OR

We will use the (specify either (1) Decay-In-Storage, or (2) Disposal of Liquids Into Sanitary Sewerage) model waste procedures that are published in Appendix T to NUREG-1556, Vol. 14, 'Program-Specific Guidance About Well Logging, Tracer, and Field Flood Study Licenses,' dated April 2000."

OR

2. "Provided are our procedures for waste collection, storage and disposal by any of the authorized methods described in this section." Applicants should contact the appropriate Regional Office of the NRC for guidance to obtain approval of any method(s) of waste disposal other than those discussed in this section.

OR

3. If access to a radioactive waste burial site is unavailable, the applicant should request authorization for extended interim storage of waste. Applicant should refer to NRC IN 90-09, "Extended Interim Storage of Low-Level Radioactive Waste by Fuel Cycle and Materials Licensees," dated February 1990 or NRC IN 93-50, "Extended Storage of Sealed Sources," dated July 1993, for guidance and submit the required information with the application.

Note: Applicants do not need to provide information to NRC if they plan to dispose of LLW via transfer to an authorized recipient. Alternative responses will be reviewed using the criteria listed above.

References: See the Notice of Availability on the inside front cover of this report to obtain copies of:

1. Policy and Guidance Directive PG 8-10, "Disposal of Incinerator Ash as Ordinary Waste," dated January 1997

2. Policy and Guidance Directive PG 94-05, "Updated Guidance on Decay-In-Storage," dated October 1994

3. Information Notice 94-23, "Guidance to Hazardous, Radioactive, and Mixed Waste Minimization Program," dated May 1994

4. Information Notice 94-07, "Solubility Criteria for Liquid Effluent Releases to Sanitary Sewerage Under the Revised 10 CFR 20," dated January 1994

5. Information Notice 84-94, "Reconcentration of Radionuclides Involving Discharges into Sanitary Sewerage Systems Permitted Under 10 CFR 20.203 (now 10 CFR 20.2003)," dated December 1984

6. Information Notice 90-09, "Extended Interim Storage of Low-Level Radioactive Waste by Fuel Cycle and Materials Licensees," dated February 1990

7. Information Notice 93-50, "Extended Storage of Sealed Sources," dated July 1993.

Information Notices are available at <http://www.nrc.gov>.

8.12   ITEM 12: FEES

The next two items on NRC Form 313 are to be completed on the form itself.

On NRC Form 313, enter the appropriate fee category from 10 CFR 170.31 and the amount of the fee enclosed with the application.

Note: Applicants who wish to perform field flood tracer studies should review 10 CFR Part 51 (particularly 10 CFR 51.30, 51.60, and 51.66) for further information concerning the environmental information needed by the NRC to prepare an environmental assessment. Environmental assessments are full-cost recovery items under 10 CFR Part 170. Full cost will be determined based on the professional staff time and appropriate staff time expended, as described in footnote e.3. to 10 CFR 170.31.

8.13   ITEM 13: CERTIFICATION

Individuals acting in a private capacity are required to date and sign NRC Form 313. Otherwise, representatives of the corporation or legal entity filing the application should date and sign

NRC Form 313. Representatives signing an application must be authorized to make binding commitments and to sign official documents on behalf of the applicant. As discussed previously in "Management Responsibility," signing the application acknowledges management's commitment and responsibilities for the radiation protection program. NRC will return all unsigned applications for proper signature.

Note:

- It is a criminal offense to make a willful false statement or representation on applications or correspondence (18 U.S.C. 1001).

- When the application references commitments, those items become part of the licensing conditions and regulatory requirements.

9   AMENDMENTS AND RENEWALS TO A LICENSE

It is the licensee's obligation to keep the license current. If any of the information provided in the original application is to be modified or changed, the licensee must submit an application for a license amendment before the change takes place. Also, to continue the license after its expiration date, the licensee must submit an application for a license renewal at least 30 days before the expiration date (10 CFR 2.109, 10 CFR 30.36(a)).

Applications for license amendment, in addition to the following, must provide the appropriate fee. For renewal and amendment requests, applicants must do the following:

Be sure to use the most recent guidance in preparing an amendment or renewal request

- Submit in duplicate, either an NRC Form 313 or a letter requesting amendment or renewal

- Provide the license number

- For renewals, provide a complete and up-to-date application if many outdated documents are referenced or there have been significant changes in regulatory requirements, NRC's guidance, the licensee's organization, or the radiation protection program. Alternatively, describe clearly the exact nature of the changes, additions, and deletions.

Using the suggested wording of responses and committing to using the model procedures in this report will expedite NRC's review.

10 APPLICATIONS FOR EXEMPTIONS

Various sections of NRC's regulations address requests for exemptions (e.g., 10 CFR 19.31, 10 CFR 20.2301, 10 CFR 30.11(a), 10 CFR 39.91). These regulations state that NRC may grant an exemption, acting on its own initiative or on an application from an interested person. Key considerations are whether the exemption is authorized by law, will endanger life or property or the common defense and security, and is otherwise in the public interest.

Until NRC has granted an exemption in writing, NRC expects strict compliance with all applicable regulations.

Exemptions are not intended for large classes of licenses, and they are generally limited to a unique situation. Exemption requests must be accompanied by descriptions of the following:

- Regulations to which the exemption is requested and why the exemption is needed

- Proposed compensatory safety measures intended to provide a level of health and safety equivalent to the regulation for which the exemption is being requested.

11 TERMINATION OF ACTIVITIES

Regulations: 10 CFR 20.1401; 10 CFR 20.1402; 10 CFR 20.1403; 10 CFR 20.1404; 10 CFR 20.1405; 10 CFR 20.1406. 10 CFR 30.34(b); 10 CFR 30.35(g); 10 CFR 30.36(d); 10 CFR 30.36(g); 10 CFR 30.36(h); 10 CFR 30.36(j); 10 CFR 30.51(f); and 10 CFR 39.91

Criteria: Pursuant to the regulations described above, the licensee must do the following:

- Notify NRC, in writing, within 60 days of:

  — the expiration of its license

  — a decision to permanently cease licensed activities at the *entire site* (regardless of contamination levels)

  — a decision to permanently cease licensed activities in *any separate building or outdoor area*, if they contain residual radioactivity making them unsuitable for release according to NRC requirements

  — no principal activities having been conducted at the *entire site* under the license for a period of 24 months

  — no principal activities having not been conducted for a period of 24 months in *any separate building or outdoor area*, if they contain residual radioactivity making them unsuitable for release according to NRC requirements.

- Submit decommissioning plan, if required by 10 CFR 30.36(g).

- Conduct decommissioning, as required by 10 CFR 30.36(h) and 10 CFR 30.36(j).

- Submit, to the appropriate NRC Regional Office, completed NRC Form 314, "Certificate of Disposition of Materials" (or equivalent information) and a demonstration that the premises are suitable for release for unrestricted use (e.g., results of final survey).

- Before a license is terminated, send the records important to decommissioning to the appropriate NRC Regional Office. If licensed activities are transferred or assigned in accordance with 10 CFR 30.34(b), transfer records important to decommissioning to the new licensee.

Discussion: As discussed above in "Criteria," before a licensee can decide whether it must notify NRC, the licensee must determine whether residual radioactivity is present and, if so, whether the levels make the building or outdoor area unsuitable for release according to NRC requirements. A licensee's determination that a facility is not contaminated is subject to verification by NRC inspection.

The permanent cessation of principal activities in an individual room or laboratory may require the licensee to notify NRC if no other licensed activities are being performed in the building.

Draft Regulatory Guide DG-4006, "Demonstrating Radiological Criteria For License Termination," issued July 8, 1998 and NUREG/BR-0241, "NMSS Handbook for

Decommissioning Fuel Cycle and Materials Licenses," dated March 1997, contain the current regulatory guidance concerning decommissioning of facilities and termination of licenses. Appendix B of the Handbook contains a comprehensive list of NRC's decommissioning regulations and guidance. NUREG-1575, "Multi-Agency Radiation Survey and Site Investigation Manual (MARSSIM)," dated December 1997, should be reviewed by licensees who have large facilities to decommission. An acceptable screening computer code for calculating screening values to demonstrate compliance with the unrestricted dose limits is D and D, Version 1; this was issued on August 20, 1998. Supplemental information on the implementation of the final rule on radiological criteria for license termination was published in the Federal Register (Volume 63, Number 222, Page 64132-64134) on November 18, 1998. This includes the following acceptable license termination screening values of common radionuclides for building surface contamination.

Table 11.1 Acceptable License Termination Screening Values of Common Radionuclides for Building Surface Contamination

Radionuclide	Symbol	Acceptable Screening Levels*
hydrogen-3 (tritium)	H-3	1.2×10^8
carbon-14	C-14	3.7×10^6
sodium-22	Na-22	9.5×10^3
sulfur -35	S-35	1.3×10^7
iron-55	Fe-55	4.5×10^6
cobalt-60	Co-60	7.1×10^3
nickel-63	Ni-63	1.8×10^6
strontium-90	Sr-90	8.7×10^6
cesium-137	Cs-137	2.8×10^4
iridium-192	Ir-192	7.4×10^4

* Screening levels are based on the assumption that the fraction of removable surface contamination is equal to 0.1. For cases when the fraction of removable contamination is undetermined or higher than 0.1, users may assume, for screening purposes, that 100% of surface contamination is removable; and therefore the screening levels should be decreased by a factor of 10. Alternatively, users having site-specific data on the fraction of removable contamination (e.g., within 10% to 100% range) may calculate site-specific screening levels using D and D Version 1, based on site-specific resuspension factor. For Unrestricted Release (dpm/100 cm²) Units are disintegrations per minute per 100 square centimeters (dpm/100 cm²). 1 dpm is equivalent to 0.0167 becquerel (Bq). The screening values represent surface concentrations of individual radionuclides that would be deemed in compliance with the 0.25 mSv/yr (25 mrem/yr) unrestricted release dose limit in 10 CFR 20.1402. For radionuclides in a mixture, the "sum of fractions" rule applies; see 10 CFR Part 20, Appendix B, Note 4. Refer to NRC Draft Guidance DG-4006 for further information on application of the values in this table.

Response from Applicant: The applicant is not required to submit a response to the NRC during the initial application. However, when the license expires or at the time the licensee ceases operations, then any necessary decommissioning activities must be undertaken, NRC Form 314 or equivalent information must be submitted, and other actions must be taken as summarized in the Criteria.

Reference: Copies of NRC Form 314, "Certificate of Disposition of Materials," are available upon request from NRC's Regional Offices. (See Figure 2.1 for addresses and telephone numbers).

Appendix A

List of Documents Considered in Development of this NUREG

List of Documents Considered in Development of this NUREG

This report incorporates and updates the guidance previously found in the NUREG reports, Regulatory Guides (RGs), Policy and Guidance Directives (P&GDs), Information Notices (INs), and Technical Assistance Requests (TARs) listed below. Other NRC documents such as Manual Chapters (MCs), Inspection Procedures (IPs), and Memoranda of Understanding (MOU) were also consulted during the preparation of this report. When this report is issued in final form, the documents marked with an asterisk (*) will be considered superseded and should not be used.

Table A.1   List of NUREG Reports, Regulatory Guides, and Policy and Guidance Directives

Document Identification	Title	Date
Working Papers		
* Working Paper	Guide for the Preparation of Applications for the Use of Radioactive Materials as Inverell Tracers in Field Flooding for the Enhanced Recovery of Oil and Natural Gas, First Draft	9/16/83
Draft Regulatory Guide		
Draft Regulatory Guide FC 413-4	Guide for the Preparation of Applications for Licenses for the Use of Radioactive Materials in Calibrating Radiation Survey and Monitoring Instruments	6/85
Draft Regulatory Guide FC 412-4	Guide for the Preparation of Applications for the Use of Radioactive Materials in Leak-Testing Services	6/85
*Draft Regulatory Guide	Guide for the Preparation of Applications for the Use of Radioactive Materials in Well Logging Operations	7/87
Regulatory Guide		
Regulatory Guide (RG) 10.8, Rev.2	Guide for the Preparation of Applications for Medical Use Programs	8/87
Regulatory Guide (RG) 3.66	Standard Format and Content of Financial Assurance Mechanisms Required for Decommissioning Under 10 CFR Parts 30, 40, 70, and 72	6/90
Regulatory Guide (RG) 4.20	Constraints on Release of Airborne Radioactive Materials to the Environment for Licensees Other Than Power Reactors	6/90
Regulatory Guide (RG) 8.7, Rev.1	Instructions for Recording and Reporting Occupational Radiation Exposure Data	6/92

Document Identification	Title	Date
Regulatory Guide (RG) 8.25	Air Sampling in the Workplace	6/92
Regulatory Guide (RG) 8.34	Monitoring Criteria and Methods to Calculate Occupational Radiation Doses	7/92
Regulatory Guide (RG) 8.9	Acceptable Concepts, Models, Equations, and Assumptions for a Bioassay Program	7/93
Regulatory Guide (RG) 8.37	ALARA Levels for Effluents from Materials Facilities	7/93
Regulatory Guide (RG) 8.32	Criteria for Establishing a Tritium Bioassay Program	7/98
NUREG		
NUREG-1541	Process and Design for Consolidating and Updating Materials Licensing Guidance	4/96
NUREG-1539	Methodology and Findings of the NRC's Materials Licensing Process Redesign	4/96
NUREG-1507	Minimum Detectable Concentrations with Typical Radiation Survey Instruments for Various Contaminants and Field Conditions	6/98
Letters		
*Generic Exemption to 10 CFR 39.41(a)(3)	All NRC Well Logging Licensee	8/10/89
SP-96-022	All Agreement States Letter	2/16/96
NCRP or ICRP Documents		
National Council on Radiation Protection and Measurements (NCRP) Report No. 49	Structural Shielding Design and Evaluation for Medical Use of X Rays and Gamma Rays of Energies Up to 10 MeV	

Document Identification	Title	Date
ANSI Documents		
ANSI N13.1	Sampling Airborne Radioactive Materials in Nuclear Facilities	1991
ANSI N323A-1997	Radiation Protection Instrumentation Test and Calibration	1997
ANSI/HPS N43.6-1997	Sealed Radioactive Sources—Classifications	1997
Other Documents		
	A Review of Department of Transportation Regulations for Transportation of Radioactive Materials (1983 revision)	
	The Health Physics & Radiological Health Handbook, Revised Edition, Edited by Bernard Shleien	1992
Technical Assistance Requests		
*Memorandum	Richard Cunningham, Subject: Proposed Abandonment of Well-Logging Source in an Artesian Well	02/05/91
*Memorandum	Bill Beach, Subject: Burial of Frac Sands as a Method of Waste Disposal	07/01/91
*Memorandum	John Glenn, Subject: Interpretation of 10 CFR 39.47 - Radioactive Markers	10/29/91
*Memorandum	John Glenn, Subject: Use of "Exempt" Sources as Well Markers	03/11/92
*Memorandum	Richard Cunningham, Subject: Well Logging Source Lost Downhole	05/22/92
*Memorandum	John Glenn, Subject: Authorization to Use Cesium-137 or Cobalt-60 Sealed Sources in Specially Designed Bullets or Core Gun Driver Assemblies for Use as Radioactive Markers in Wells	05/17/93
*Memorandum	John Glenn, Subject: Exemption from Semiannual Timer Period for Equipment Inspection and Maintenance Specified in 10 CFR 39.43(b) When Equipment is in Storage	02/24/94
*Memorandum	John Glenn, Subject: Temporary Exemption From 10 CFR 39.41	03/17/94

Document Identification	Title	Date
*Memorandum	Carl Paperiello, Subject: Use of Tritiated Hexadecane as a Liquid for the Purpose of Tagging a Hydrocarbon-Based Gel to be Used in Tagging Mud Used in Oil Well Drilling Applications	04/13/94
*Memorandum	Carl J. Paperiello, Subject: Exemption to 10 CFR 39.47 Markers, 39.35(c) Leak Testing, 39.15(a)(3) Agreements, and 39.77(c)(1) Notification of Lost Source	09/21/94
*Memorandum	John Glenn, Subject: Exemption to 10 CFR 39.47 Markers, 39.35(c) Leak Testing, 39.15(a)(3) Agreements, and 39.77(c)(1) Notification of Lost Source	09/21/94
*Memorandum	Larry Camper, Subject: Use of Handling Tools When Using Tracer Materials	06/12/95
*Memorandum	Larry Camper, Subject: Request an Amendment to License to Add a Well as a Storage Site for a PDK Logging Tool	06/14/95
*Memorandum	Larry Camper, Subject: Alternative Training for Well Logging Supervisor	06/26/95
Information Notices		
IN 90-09	Extended Interim Storage of Low-Level Radioactive Waste by Fuel Cycle and Material Licensees	2/90
IN 93-50	Extended Storage of Sealed Sources	7/8/93
IN 89-25 (Rev. 1)	Unauthorized Transfer of Ownership or Control of Licensed Activities	12/7/94
IN 94-07	Solubility Criteria for Liquid Effluent Releases to Sanitary Sewerage Under the Revised 10 CFR 20	2/94
IN 94-23	Guidance to Hazardous, Radioactive, and Mixed Waste Minimization Program	3/94
IN 96-28	Suggested Guidance Relating to Development and Implementation of Corrective Action	5/96
IN 97-30	Control of Licensed Material During Reorganizations, Employee-Management Disagreements and Financial Crises	6/97

Document Identification	Title	Date
Policy Guidance and Directives		
Policy and Guidance Directive FC 90-2 (Rev. 1)	Standard Review Plan for Evaluating Compliance with Decommissioning Requirements	4/30/91
Revision 1, Supplement to Policy and Guidance Directive FC 84-20	Impact of Revision of 10 CFR Part 51 on Materials License Actions	3/94
Policy and Guidance Directive PG 8-11	NMSS Procedures for Reviewing Declarations of Bankruptcy	8/8/96
Inspection Procedures		
Inspection Procedure (IP) 87103	Inspection of Material Licensees Involved in an Incident or Bankruptcy Filing	2/97
Inspection Procedure (IP) 87113	Appendix A - "Well Logging Inspection Field Notes"	1998

Appendix B

United States Nuclear Regulatory Commission Form 313

<table>
<tr><td colspan="2">NRC FORM 313
(8-1999)
10 CFR 30, 32, 33
34, 35, 36, 39 and 40

APPLICATION FOR MATERIAL LICENSE</td><td>U. S. NUCLEAR REGULATORY COMMISSION</td><td>APPROVED BY OMB: NO. 3150-0120   EXPIRES:08/31/2002

Estimated burden per response to comply with this mandatory information collection request: 7.4 hours. Submittal of the application is necessary to determine that the applicant is qualified and that adequate procedures exist to protect the public health and safety. Send comments regarding burden estimate to the Records Management Branch (T-6 E6), U.S. Nuclear Regulatory Commission, Washington, DC 20555-0001, or by internet e-mail to bjs1@nrc.gov, and to the Desk Officer, Office of Information and Regulatory Affairs, NEOB-10202, (3150-0120), Office of Management and Budget, Washington, DC 20503. If a means used to impose an information collection does not display a currently valid OMB control number, NRC may not conduct or sponsor, and a person is not required to respond to, the information collection</td></tr>
</table>

INSTRUCTIONS: SEE THE APPROPRIATE LICENSE APPLICATION GUIDE FOR DETAILED INSTRUCTIONS FOR COMPLETING APPLICATION. SEND TWO COPIES OF THE ENTIRE COMPLETED APPLICATION TO THE NRC OFFICE SPECIFIED BELOW.

APPLICATION FOR DISTRIBUTION OF EXEMPT PRODUCTS FILE APPLICATIONS WITH:

DIVISION OF INDUSTRIAL AND MEDICAL NUCLEAR SAFETY
OFFICE OF NUCLEAR MATERIALS SAFETY AND SAFEGUARDS
U.S. NUCLEAR REGULATORY COMMISSION
WASHINGTON, DC 20555-0001

ALL OTHER PERSONS FILE APPLICATIONS AS FOLLOWS:

IF YOU ARE LOCATED IN:

CONNECTICUT, DELAWARE, DISTRICT OF COLUMBIA, MAINE, MARYLAND, MASSACHUSETTS, NEW HAMPSHIRE, NEW JERSEY, NEW YORK, PENNSYLVANIA, RHODE ISLAND, OR VERMONT, SEND APPLICATIONS TO:

LICENSING ASSISTANT SECTION
NUCLEAR MATERIALS SAFETY BRANCH
U.S. NUCLEAR REGULATORY COMMISSION, REGION I
475 ALLENDALE ROAD
KING OF PRUSSIA, PA 19406-1415

ALABAMA, FLORIDA, GEORGIA, KENTUCKY, MISSISSIPPI, NORTH CAROLINA, PUERTO RICO, SOUTH CAROLINA, TENNESSEE, VIRGINIA, VIRGIN ISLANDS, OR WEST VIRGINIA, SEND APPLICATIONS TO:

SAM NUNN ATLANTA FEDERAL CENTER
U.S. NUCLEAR REGULATORY COMMISSION, REGION II
61 FORSYTH STREET, S.W., SUITE 23T85
ATLANTA, GEORGIA 30303-8931

IF YOU ARE LOCATED IN:

ILLINOIS, INDIANA, IOWA, MICHIGAN, MINNESOTA, MISSOURI, OHIO, OR WISCONSIN, SEND APPLICATIONS TO:

MATERIALS LICENSING SECTION
U.S. NUCLEAR REGULATORY COMMISSION, REGION III
801 WARRENVILLE RD.
LISLE, IL 60532-4351

ALASKA, ARIZONA, ARKANSAS, CALIFORNIA, COLORADO, HAWAII, IDAHO, KANSAS, LOUISIANA, MONTANA, NEBRASKA, NEVADA, NEW MEXICO, NORTH DAKOTA, OKLAHOMA, OREGON, PACIFIC TRUST TERRITORIES, SOUTH DAKOTA, TEXAS, UTAH, WASHINGTON, OR WYOMING, SEND APPLICATIONS TO:

NUCLEAR MATERIALS LICENSING SECTION
U.S. NUCLEAR REGULATORY COMMISSION, REGION IV
611 RYAN PLAZA DRIVE, SUITE 400
ARLINGTON, TX 76011-8064

PERSONS LOCATED IN AGREEMENT STATES SEND APPLICATIONS TO THE U.S. NUCLEAR REGULATORY COMMISSION ONLY IF THEY WISH TO POSSESS AND USE LICENSED MATERIAL IN STATES SUBJECT TO U.S. NUCLEAR REGULATORY COMMISSION JURISDICTIONS.

1. THIS IS AN APPLICATION FOR (Check appropriate item)
   - [ ] A. NEW LICENSE
   - [ ] B. AMENDMENT TO LICENSE NUMBER ____
   - [ ] C. RENEWAL OF LICENSE NUMBER ____

2. NAME AND MAILING ADDRESS OF APPLICANT (Include Zip code)

3. ADDRESS(ES) WHERE LICENSED MATERIAL WILL BE USED OR POSSESSED

4. NAME OF PERSON TO BE CONTACTED ABOUT THIS APPLICATION
   TELEPHONE NUMBER

SUBMIT ITEMS 5 THROUGH 11 ON 8-1/2 X 11" PAPER. THE TYPE AND SCOPE OF INFORMATION TO BE PROVIDED IS DESCRIBED IN THE LICENSE APPLICATION GUIDE.

5. RADIOACTIVE MATERIAL. a. Element and mass number; b. chemical and/or physical form; and c. maximum amount which will be possessed at any one time.

6. PURPOSE(S) FOR WHICH LICENSED MATERIAL WILL BE USED.

7. INDIVIDUAL(S) RESPONSIBLE FOR RADIATION SAFETY PROGRAM AND THEIR TRAINING EXPERIENCE.

8. TRAINING FOR INDIVIDUALS WORKING IN OR FREQUENTING RESTRICTED AREAS.

9. FACILITIES AND EQUIPMENT.

10. RADIATION SAFETY PROGRAM.

11. WASTE MANAGEMENT.

12. LICENSEE FEES (See 10 CFR 170 and Section 170.31)
   FEE CATEGORY   AMOUNT ENCLOSED $

13. CERTIFICATION. (Must be completed by applicant) THE APPLICANT UNDERSTANDS THAT ALL STATEMENTS AND REPRESENTATIONS MADE IN THIS APPLICATION ARE BINDING UPON THE APPLICANT.

THE APPLICANT AND ANY OFFICIAL EXECUTING THIS CERTIFICATION ON BEHALF OF THE APPLICANT, NAMED IN ITEM 2, CERTIFY THAT THIS APPLICATION IS PREPARED IN CONFORMITY WITH TITLE 10, CODE OF FEDERAL REGULATIONS, PARTS 30, 32, 33, 34, 35, 36, 39 AND 40, AND THAT ALL INFORMATION CONTAINED HEREIN IS TRUE AND CORRECT TO THE BEST OF THEIR KNOWLEDGE AND BELIEF

WARNING: 18 U.S.C. SECTION 1001 ACT OF JUNE 25, 1948 62 STAT. 749 MAKES IT A CRIMINAL OFFENSE TO MAKE A WILLFULLY FALSE STATEMENT OR REPRESENTATION TO ANY DEPARTMENT OR AGENCY OF THE UNITED STATES AS TO ANY MATTER WITHIN ITS JURISDICTION.

CERTIFYING OFFICER – TYPED/PRINTED NAME AND TITLE   SIGNATURE   DATE

FOR NRC USE ONLY

TYPE OF FEE	FEE LOG	FEE CATEGORY	AMOUNT RECEIVED $	CHECK NUMBER	COMMENTS
APPROVED BY				DATE	

Appendix C

Suggested Format for Providing Information Requested in Items 5 through 11 of NRC Form 313

Suggested Format for Providing Information Requested in Items 5 through 11 of NRC Form 313

Item No.	Title and Criteria	Use Table Below	Description Attached
5	**RADIOACTIVE MATERIAL** **Sealed Sources and Devices** • Identify each radionuclide that will be used in sealed sources	[ ]	[ ]
	• Identify each radionuclide that will be used in energy compensation sources	[ ]	[ ]
	• Identify each radionuclide that will be used as tracer materials in single wells	[ ]	[ ]
	• Identify each radionuclide that will be used as tracer materials in field flood studies in multiple wells	[ ]	[ ]
	• Identify any depleted uranium that is used as shielding material or sinker bars.	[ ]	[ ]

Well Logging Sealed Sources		
Radioisotope	**Manufacturer/Model No.**	**Quantity**
		Not to exceed the maximum activity per source as specified in the Sealed Source and Device Registration Sheet.
		Not to exceed the maximum activity per source as specified in the Sealed Source and Device Registration Sheet.
		Not to exceed the maximum activity per source as specified in the Sealed Source and Device Registration Sheet.

Neutron Generators		
Radioisotope	**Manufacturer/Model No.**	**Quantity**

Electronic Compensation Sources

Radioisotope	Manufacturer/Model No.	Quantity
		Not to exceed the maximum activity per source as specified in the Sealed Source and Device Registration Sheet.
		Not to exceed the maximum activity per source as specified in the Sealed Source and Device Registration Sheet.

Tracer Materials

Radioisotope	Chemical or Physical Form			Millicuries Per Injection	Total Quantity Requested
	[ ] Gas	[ ] Liquid	[ ] Labeled Frac Sands		
	[ ] Gas	[ ] Liquid	[ ] Labeled Frac Sands		
	[ ] Gas	[ ] Liquid	[ ] Labeled Frac Sands		

Depleted Uranium

Radioisotope	Manufacturer/Model No.	Kilograms Requested
Depleted Uranium (DU)		

Sealed Sources Not Used in Well Logging Operations

Radioisotope	Manufacturer/Model No.	Quantity
		Not to exceed the maximum activity per source as specified in the Sealed Source and Device Registration Sheet.

Radioisotope	Manufacturer/Model No.	Quantity
		Not to exceed the maximum activity per source as specified in the Sealed Source and Device Registration Sheet.

Commitment:	Yes	N/A
Confirm that each sealed source used in above ground devices is registered as an approved sealed source or device by NRC or an Agreement State and will be possessed and used in accordance with the conditions specified in the registration certificate.	[ ]	[ ]

Item No.	Title and Criteria	Yes	N/A	Description Attached
	RADIOACTIVE MATERIAL **Financial Assurance and Record Keeping for Decommissioning** • Pursuant to 10 CFR 30.35(g), we shall maintain drawings and records important to decommissioning and transfer these records to a new licensee before licensed activities are transferred, or assign the records to the appropriate NRC Regional Office before the license is terminated. **OR** • If financial assurance is required, submit evidence.	[ ] [ ]	[ ] [ ]	 [ ]

Item No.	Title and Criteria	Yes	N/A	Description Attached
6	**PURPOSE(S) FOR WHICH LICENSED MATERIAL WILL BE USED**			
	• Oil and Gas Well Logging.	[ ]	[ ]	
	• Mineral Well Logging.	[ ]	[ ]	
	• Geophysical Well Logging.	[ ]	[ ]	
	• Tracer Studies in Single Wells.	[ ]	[ ]	
	• Field Flood or Enhanced Recovery Studies in Multiple Wells.	[ ]	[ ]	
	OR			
	• Specify the purposes for which the sources and device(s) will be used other than those included in the manufacturer's recommendations, and as specified on the SSD Registration Certificate.			[ ]
	AND			
	• We plan to perform in *fresh water* aquifers:			
	– Tracer Studies	[ ]	[ ]	
	– Well logging using sealed sources	[ ]	[ ]	
	– Well logging using neutron generator.	[ ]	[ ]	

Item No.	Title and Criteria	Yes	N/A	Description Attached
7	**INDIVIDUAL(S) RESPONSIBLE FOR RADIATION SAFETY PROGRAM AND THEIR TRAINING EXPERIENCE** **Radiation Safety Officer (RSO)** • The name of the proposed RSO and other individuals who will be responsible for the radiation protection program. 　　Name:_____	[ ]		
	• Demonstrate that the RSO has sufficient independence and direct communication with responsible management officials by providing a copy of an organizational chart by position, demonstrating day-to-day oversight of the radiation safety activities			[ ]
	AND EITHER			
	• The specific training and experience of the RSO			[ ]
	OR			
	• Alternative information demonstrating that the proposed RSO is qualified by training and experience, e.g., listed by name as an authorized user or the RSO on an NRC or Agreement State license that requires a radiation safety program of comparable size and scope.		[ ]	[ ]

Item No.	Title and Criteria	Yes	N/A	Description Attached
8	**TRAINING FOR LOGGING SUPERVISORS AND LOGGING ASSISTANTS**			
	• Submit an outline of the training to be given to prospective logging supervisors and logging assistants.			[ ]
	• Submit your procedures for experienced logging supervisors who have worked for another licensee.			[ ]
	• Provide a copy of a typical examination and the correct answers to the examination questions. State the passing grade %.			[ ]
	• Specify the qualifications of your instructors.			[ ]
	• If training will be conducted by someone outside the applicant's organization, identify the course by title and provide the name and address of the company providing the training.	[ ]	[ ]	[ ]
	• Describe the field (practical) examination that will be given to prospective logging supervisors and logging assistants.			[ ]
	• Describe the annual refresher training program, including topics to be covered and how the training will be conducted.			[ ]
	• Submit a description of your program for inspecting the job performance of each well logging supervisor or logging assistant at intervals not to exceed 12 months, as described in 10 CFR 39.13.			[ ]

Item No.	Title and Criteria	Yes	N/A	Description Attached
9	**FACILITIES AND EQUIPMENT**			
	• Submit a drawing or sketch of the proposed facility, identifying areas where radioactive materials, including radioactive wastes, will be used or stored.		[ ]	[ ]
	• Drawings should show, where applicable, adjacent buildings, boundary lines, security fences, and lockable storage areas.		[ ]	[ ]
	• Illustrate area(s) where explosive, flammable, or other hazardous materials may be stored.		[ ]	[ ]
	• Drawings should also show the relationship and distance between restricted areas and adjacent unrestricted areas.		[ ]	[ ]
	• Drawings should specify shielding materials (concrete, lead, etc.) and means for securing radioactive materials from unauthorized removal.		[ ]	[ ]
	• Submit a drawing or sketch of the proposed tracer material storage facilities, including rooms, buildings, below ground bunker storage areas, or containers used for storage of both tracer and tracer waste materials, if appropriate. Specify the types and amount of shielding materials (concrete, lead, etc.) and means for securing tracer materials from unauthorized removal.		[ ]	[ ] [ ]
	• Describe protective clothing (such as rubber gloves, coveralls, respirators, and face shields), auxiliary shielding, absorbent materials, injection equipment, secondary containers for waste water storage for decontamination purposes, plastic bags for storing contaminated items, etc. that will be available at well sites when using tracer materials.		[ ]	[ ]
	• Describe proposed laundry facilities, if applicable, used for contaminated protective clothing. Specify how the contaminated waste water from the laundry machines or sinks is disposed. Operating and emergency procedures should address decontamination of the laundry area and equipment.		[ ]	[ ]
	• Describe proposed decontamination facilities for trucks, tracer injection tools, or other equipment contaminated by tracer materials, if applicable. Specify how the contaminated waste water for these decontamination facilities is disposed. Operating and emergency procedures should address decontamination of these types of equipment and facilities.		[ ]	[ ] [ ]

Item No.	Title and Criteria	Yes	N/A	Description Attached
9	**FACILITIES AND EQUIPMENT** *(Cont'd)*			
	• Describe, if applicable, equipment for "repackaging" gaseous, volatile, or finely divided tracer material. Most tracer users do not repackage materials and acquire their injections in precalibrated amounts or "ready to use" forms. However, should an applicant request the ability to repackage tracer, volatile, or finely divided material, the following equipment should be considered when repackaging tracer materials: sinks, trays with absorbent material, glove boxes, fume hoods with charcoal filtration, filtered exhaust, special handling equipment including special tools, rubber gloves, etc.		[ ]	[ ]

Item No.	Title and Criteria	Yes	N/A	Description Attached
10	**RADIATION SAFETY PROGRAM** The applicant is required to establish and submit its radiation protection program. The format use for providing information should be developed by the applicant. No specific format is required by NRC for submitting a radiation safety program.			[ ]
	Radiation Safety Program Audit: The applicant is *not* required to, and should not, submit its audit program to the NRC for review during the licensing phase.	Need Not Be Submitted With Application		
	Well Owner Operator/Agreement			[ ]
	Instruments • A description of the instrumentation (as described above) that will be used to perform required surveys. **OR**			[ ]
	• We will use instruments that meet the radiation monitoring instrument specifications published in Appendix N to NUREG-1556, Vol. 14, 'Program-Specific Guidance About Well Logging, Tracer and Field Flood Studies,' dated May 2000. **AND**	[ ]		
	• We will implement the model survey meter calibration program published in Appendix N to NUREG-1556, Vol. 14, 'Program-Specific Guidance About Well Logging, Tracer and Field Flood Studies,' dated May 2000. We reserve the right to upgrade our survey instruments as necessary. **OR**	[ ]	[ ]	
	• A description of alternative equipment and/or procedures for ensuring that appropriate radiation monitoring equipment will be used during licensed activities and that proper calibration and calibration frequency of survey equipment will be performed. Further, the statement "We reserve the right to upgrade our survey instruments as necessary" should be added to the response.		[ ]	[ ]

Item No.	Title and Criteria	Yes	N/A	Description Attached
10	**RADIATION SAFETY PROGRAM** *(Cont'd)*			
	Material Receipt and Accountability			
	• Physical inventories will be conducted and documented at intervals not to exceed six months, to account for all byproduct materials (sealed sources and tracer materials) and devices containing depleted uranium received and possessed under the license.	[ ]		
	Occupational Dosimetry			
	• Film badge, TLD, or OSL dosimeter will be processed and evaluated by a NVLAP-accredited entity, exchanged at the approved frequency, and worn by well logging supervisors and logging assistants.	[ ]	[ ]	
	<div align="center">**AND/OR**</div>			
	• Individual logging supervisors and logging assistants using more than 50 millicuries of iodine-131 at any one time or in any 5-day period will be provided a bioassay.	[ ]	[ ]	
	• Bioassay plan attached.			[ ]
	• Individual logging supervisors and logging assistants will not use more than 50 millicuries of iodine-131 at any one time or in any 5-day period at field stations or at temporary job sites.	[ ]	[ ]	
	• We will contract with an outside group for bioassay services.	[ ]	[ ]	
	• Each vendor is licensed or otherwise authorized by NRC or an Agreement State to provide required bioassay services.	[ ]	[ ]	
	Public Dose The applicant is not required to, and should not, submit a response to the public dose section during the licensing phase. This matter will be inspected during an inspection.	Need Not Be Submitted With Application		

Item No.	Title and Criteria	Yes	N/A	Description Attached
10	**RADIATION SAFETY PROGRAM** *(Cont'd)*			
	Leak Tests			
	• Leak tests, when required by the license, will be performed at intervals approved by the NRC or an Agreement State and specified in the Sealed Source and Device Registration Sheet. Leak tests will be performed either by an organization authorized by NRC or an Agreement State to provide leak testing services to other licensees or using a leak test kit supplied by an organization authorized by NRC or an Agreement State to provide leak test kits to other licensees and according to the kit supplier's instructions.	[ ]		
	• Leak testing and analysis will be done by the applicant, and the information in Appendix R supporting a request to perform leak testing and sample analysis is attached.	[ ]		[ ] [ ]
	• We will follow alternate procedures, and our specific procedures are enclosed for review.	[ ]	[ ]	[ ]
	Daily Maintenance			
	• A description of procedure(s) for conducting daily visual inspection is submitted.			[ ]
	OR			
	• Visual daily inspections will be conducted and records maintained in accordance with Section 8.10.9.1 of NUREG-1556, Vol. 14 to ensure that well logging equipment is in good working condition and that required labeling is present.	[ ]	[ ]	

Item No.	Title and Criteria	Yes	N/A	Description Attached
10	**RADIATION SAFETY PROGRAM** *(Cont'd)*			
	Semi-Annual Maintenance			
	• Procedure(s) for conducting semi-annual inspections and routine maintenance of source holders, logging tools, injection tools, source handling tools, storage containers, transport containers, and uranium sinker bars to ensure that the labeling required by 10 CFR Part 39 is legible and that no physical damage is visible, is attached.			[ ]
	OR			
	• Semi-annual inspections and routine maintenance will be conducted and records maintained for source holders, logging tools, injection tools, source handling tools, storage containers, transport containers, and uranium sinker bars in accordance with Section 8.10.9.2 of NUREG 1556, Vol. 14, to ensure that well logging equipment is in good working condition with no physical damage evident and that the required labeling is present.	[ ]	[ ]	
	Maintenance Requiring Special Authorization			
	• Prohibited activities described in Section 8.10.9.3 of NUREG-1556, Vol. 14 will not be conducted unless approved by the NRC.	[ ]	[ ]	
	OR			
	• Detailed procedures for any prohibited activities, including radiation safety precautions that individuals will be expected to follow when performing these tasks and the minimum qualifications of these individuals, are attached. Each different task must is. Should a procedure require the removal of the sealed source from the holder before performing any maintenance on the holder, applicants should describe the removal procedures.		[ ]	[ ]
	Transportation No response is needed from applicants during the licensing phase. Transportation issues are reviewed during inspections.	No Response is Necessary for this Section		

Item No.	Title and Criteria	Yes	N/A	Description Attached
10	**RADIATION SAFETY PROGRAM** *(Cont'd)*	No Response is Necessary for this Section		
	Minimization of Contamination			
	The applicant does not need to provide a response to this item under the following conditions, and NRC will consider that the above criteria have been met if the applicant's responses meet the criteria in the following sections: "Facilities and Equipment," "Radiation Safety Program - Tracer Studies," "Radiation Safety Program - Operating and Emergency Procedures," and "Radiation Safety Program - Waste Management."			
	AND			
	• Major decontamination procedures *will not be performed.* Decontamination of the facilities or sealed sources require special authorization from the NRC or an Agreement State.	[ ]	[ ]	
	OR			
	• Major decontamination procedures *will be performed*, and procedures to perform major decontamination activities are provided. Applicants should submit their procedures to perform major decontamination activities if they intend to perform the activity rather than contracting the work to a licensed entity.	[ ]	[ ]	[ ]
	Drill-to-stop			
	• Operating and emergency procedures for conducting DTS well logging operations submitted.	[ ]	[ ]	[ ]
	OR			
	• A summary addressing important radiation safety aspects of its O&E Procedures when conducting DTS submitted.	[ ]	[ ]	[ ]

Item No.	Title and Criteria	Yes	N/A	Description Attached
10	**RADIATION SAFETY PROGRAM** *(Cont'd)*			
	Measurement While Drilling or Logging While Drilling			
	• Operating and emergency procedures for conducting MWD and/or LWD well logging operations submitted.	[ ]	[ ]	[ ]
	OR			
	• Summary that addresses important radiation safety aspects of Operating and Emergency Procedures when conducting MWD and/or LWD well logging operations submitted.			[ ]
	Energy Compensation Sources			
	• Operating and emergency procedures for using ECDs submitted.	[ ]	[ ]	[ ]
	OR			
	• A summary or outline addressing important radiation safety aspects of operating and emergency procedures when using or handling ECSs submitted.	[ ]	[ ]	[ ]
	- Instructions for testing ECSs requiring leak tests at intervals not to exceed 3 years	[ ]		
	- Instructions for conducting physical inventories of ECSs at least every 6 months	[ ]		
	- A record system for maintaining inventory records required by 10 CFR 39.37	[ ]		
	- A record system for maintaining records of use for ECSs.	[ ]		
	Use of Sealed Sources or Neutron Generators in Fresh Water Aquifers	No response is required from the licensee unless it requests authorization for the prohibited activity.		
	Tracer Studies in Single Well Applications	No response required to this section provided that the elements listed in 8.10.13.1 are contained in other sections.		

Item No.	Title and Criteria	Yes	N/A	Description Attached
10	**RADIATION SAFETY PROGRAM** *(Cont'd)*			
	Field Flood and Secondary Recovery Applications (Tracer Studies in Multiple Wells)			
	• We will be using tracer materials in conducting field flood studies in multiple wells.	[ ]		
	• We will not conduct field flood studies.	[ ]		
	OR			
	• We have submitted the information outlined in Appendix F for conducting field flood studies.	[ ]	[ ]	[ ]
	Tracer Studies in Fresh Water Aquifers			
	• We will not knowingly inject tracer material into a fresh water aquifer.	[ ]		
	OR			
	• Applicants requesting authorization to inject licensed radioactive material into a fresh aquifer must provide their reasons for performing the study and procedures to protect their workers and the public. Licensees must also provide the information required for an environmental assessment. Authorization to conduct such activities requires that applicants provide procedures to safeguard the public, licensee personnel, and the environment, in addition to providing an environmental impact study.	[ ]	[ ]	[ ]

Item No.	Title and Criteria	Yes	N/A	Description Attached
10	**RADIATION SAFETY PROGRAM** *(Cont'd)*			
	Radioactive Collar and Subsidence or Depth Control Markers			
	• We will only use radioactive markers where each individual marker contains only quantities of licensed material not exceeding the exempt quantities authorized in 10 CFR 30.71, Schedule B, as described in Section 8.10.14 of NUREG-1556, Vol. 14.	[ ]	[ ]	
	OR			
	• Procedures for using radioactive markers that are in excess of the quantities in Section 8.10.14 of NUREG-1556, Vol.14. are submitted for review.		[ ]	[ ]
	Neutron Accelerators Using Licensed Material			
	• We will not use neutron generators (accelerators) in our well logging operations.	[ ]		
	OR			
	• We will use neutron generators (accelerators) in accordance with the criteria in Section 8.10.15 of NUREG-1556, Vol. 14.	[ ]		

Item No.	Title and Criteria	Yes	N/A	Description Attached
10	**RADIATION SAFETY PROGRAM** *(Cont'd)*			
	Depleted Uranium Sinker Bars			
	• Depleted uranium sinker bars will be obtained under the provisions of a general license, per 10 CFR 40.51, and registration form NRC Form 244 will be filed, as required.	[ ]	[ ]	
	OR			
	• Depleted uranium sinker bars will not be obtained under the provision of a general license per 10 CFR 40.51 (general license).	[ ]	[ ]	
	AND			
	• Uranium sinker bars will be possessed and inspected as specified in Section 8.10.16 of NUREG-1556, Vol. 14.	[ ]	[ ]	
	AND			
	• We have specified the number of kilograms of specifically licensed source material (DU) that should be included in the license.	[ ]	[ ]	

Item No.	Title and Criteria	Yes	N/A	Description Attached
10	**RADIATION SAFETY PROGRAM** (*Cont'd*)			
	Waste Management			
	• We will use the model waste procedures published in Appendix T to NUREG-1556, Vol. 14, "Program-Specific Guidance About Well Logging, Tracer, and Field Flood Study Licenses," dated May 2000.	[ ]	[ ]	
	OR			
	• "We will use the (specify either (1) Decay-In-Storage, or (2) Disposal of Liquids Into Sanitary Sewerage) model waste procedures that are published in Appendix T to NUREG-1556, Vol. 14, "Program-Specific Guidance About Well Logging, Tracer, and Field Flood Study Licenses," dated May 2000.	[ ]	[ ]	[ ]
	OR			
	• Provided are our procedures for waste collection, storage and disposal by any of the authorized methods described in this section. Applicants should contact the appropriate Regional Office of the NRC for guidance to obtain approval of any method(s) of waste disposal other than those discussed in this section.		[ ]	[ ]
	OR			
	• If access to a radioactive waste burial site is unavailable, the applicant should request authorization for extended interim storage of waste. Applicant should refer to NRC IN 90-09, "Extended Interim Storage of Low-Level Radioactive Waste by Fuel Cycle and Materials Licensees," dated February 1990, for guidance and submit the required information with the application.	[ ]	[ ]	[ ]

Appendix D

Checklist for License Application

Checklist for License Application

WELL LOGGING APPLICATION
REVIEW CHECK LIST

Date: _____

CONTENTS OF APPLICATION

ITEM 8.1   **TYPE OF APPLICANT/LICENSEE**

Type of Action	License No.
[ ] A. New License	Not Applicable
[ ] B. Amendment	
[ ] C. Renewal	

ITEM 8.2   **NAME OF APPLICANT/LICENSEE**

LEGAL NAME: _____

MAILING ADDRESS: _____

ITEM 8.3   **LOCATION OF USE**

[ ]   Address listed above

[ ]   Field Stations (Street Address, City, State, and Zip Code):

     ( ) _____

     ( ) _____

     ( ) _____

[ ]   Temporary Job Sites

[ ]   See attached list

ITEM 8.4   **CONTACT PERSON**

NAME: _____

TELEPHONE NUMBER: _____

ITEMS 8.5 - 8.6   **RADIOACTIVE MATERIAL TO BE POSSESSED/*REQUESTED USE OF MATERIALS**

[ ]   Energy Compensation Sources (ECS)

[ ]   Tracer Materials

[ ]   Well Logging Sealed Sources (MWD/LWD/DTS)

[ ]   Radioactive Collar/Subsidence/Depth Markers

[ ]   Depleted Uranium

[ ]   Neutron Accelerator Targets

[ ]   Sealed Sources for use above ground for other than well logging applications

SEALED MATERIALS

[ ]   Identify each radionuclide (element name and mass number) that will be used in each sealed source.

[ ]   Provide the manufacturer's (distributor's) name and model number for each sealed source and, if applicable, device requested.

[ ]   Confirm that the activity per source and maximum activity in each device will not exceed the maximum activity listed on the approved certificate of registration issued by NRC or by an Agreement State.

[ ]   Confirm that each sealed source, device, and source/device combination is registered as an approved sealed source or device by NRC or an Agreement State.

[ ]   Sealed sources that were manufactured before July 14, 1989 may use either the design and performance criteria from the United States of America Standards Institute (USASI) N5 10-1968 (10 CFR 39.41(b)) or the criteria specified in 10 CFR 39.41 or the requirements in 10 CFR 39.41(a) (1) and (2), and ANSI/HPS N43.6-1997, "Sealed Radioactive Sources - Classification." The use of the USASI standard is based on an NRC Notice of Generic Exemption issued on July 25, 1989 (54 FR 30883). See Appendix J.

[ ]   Sealed sources manufactured after July 14, 1989, are required to satisfy the requirements of 10 CFR 39.41 or the requirements in 10 CFR 39.41(a) (1) and (2) and ANSI/HPS N43.6-1997.

UNSEALED TRACER MATERIAL (Volatile & Nonvolatile)

[ ]   Provide element name with mass number, chemical and/or physical form, and maximum requested possession limit.

[ ]   Provide information for *volatile materials*, if known, on the anticipated rate of volatility or dispersion. This information may be obtained from the tracer material vendor, supplier, or manufacturer.

SEALED SOURCES

Radioisotope	Mfg./Model No. SSD Certificate No.	Quantity (Curies/MBq/GBq)	*Use

UNSEALED TRACER MATERIALS

Radioisotope	-Chemical/Physical Form -Max. Amount Used Per Injection	Quantity (Curies/MBq/GBq)	*Use	Volatility/Dispersion

*MATERIAL USE LEGEND

O=Oil Well Logging	G=Gas Well Logging	M=Mineral Well Logging	T=Tracer Studies in single wells
FF=Field Flood or Enhanced Recovery Operations	N=Neutron Generators	C=Calibration Sources in above ground applications	

FINANCIAL ASSURANCE

[ ] Financial Assurance not required

[ ] Financial Assurance required and Financial Assurance Documents submitted for review

[ ] Decommissioning Records File established

ITEM 8.7   RESPONSIBLE INDIVIDUALS

Corporate Organization Chart Submitted for Review:

Radiation Safety Organizational Chart Submitted for Review:

Name(s) of responsible individual(s)

[ ] Title(s) of individual(s)

[ ] Training of individual(s)

[ ] Experience (1-year minimum)

Radiation Safety Officer Information:

[ ] Name	[ ] Experience	[ ] Training	[ ] Independent Authority to stop unsafe operations
[ ] Organizational Chart (Day-to-Day Radiation Safety Positions) provided			
[ ] Alternative Training and Experience, if applicable			
[ ] See Appendix for the minimum RSO duties and responsibilities			

ITEM 8.8   TRAINING FOR WELL LOGGING SUPERVISORS AND WELL LOGGING ASSISTANTS

LOGGING ASSISTANT TRAINING [§39.61(b)] and (d)

[ ] In-house Training:

    ( ) Received copies of Parts 19, 20, & OE Procedures

    ( ) Classroom instruction in Parts 19 & 20 (2-4 hours)

    ( ) Instruction in the use of licensed materials, remote handling tools, survey equipment, etc. (1-2 hours)

    ( ) Successfully completed a verbal or written examination

    ( ) Exam with key
        Minimum passing grade _____ %

    ( ) Records maintained for 3 years (copies of quizzes and dates of oral examinations)

LOGGING SUPERVISOR TRAINING [§39.61(a) and (d)]

[ ] §39.61(e) Topics, by vendor
    Vendor(s) Name: _____

[ ] Instructor's Name: _____

[ ] Instructor's Qualifications:

[ ] Classroom Training Conducted by Licensee (~24 hours):

    ( ) §39.61(e) Topics:

        < > Fundamentals of radiation safety

        < > Characteristics of radiation

        < > Units of radiation dose and quantity of radiation

        < > Hazard of exposure

        < > Levels of radiation for licensed material

        < > Methods of controlling radiation dose (time, distance, shielding)

        < > Radiation safety practices, including prevention of contamination, and methods of decontamination

    ( ) Radiation detection instruments:

        < > Use

        < > Operation

        < > Calibration

        < > Instrument limitations

        < > Survey techniques

        < > Use of personnel monitoring

    ( ) Equipment to be used:

        < > Operation of equipment, including:

            □ source handling equipment

            □ remote handling tools

        < > Storage, control, and disposal of licensed material

        < > Maintenance of equipment

    ( ) Federal regulations

    ( ) Case histories

[ ] In-house Classroom Training (~8 hours):

    ( ) 10 CFR 19, 20, & 39

    ( ) OE Procedures (§39.63)

    ( ) License

    ( ) ~8 hours of classroom instruction in the above

    ( ) Successfully completed a written examination
Minimum passing grade _____ %
Exam Key

    ( ) In-house instructor qualifications

    ( ) Maintain for 3 years copies of written quizzes

( ) Field training

( ) Field/practical exam

[ ] On-the-job Training:

( ) 3 months (520 hours)

( ) 1 month (160 hours) mineral well logging

( ) 50 tracer operations or 3 months OJT

[ ] Logging supervisors with previous training

ALTERNATIVE TO DESCRIBING A TRAINING PROGRAM

[ ] Identify each individual to be specified on the license as logging supervisor or logging assistant

[ ] For each individual identified, provide the following:

( ) Copies of graded tests

( ) Certificate of course completion

( ) Details of previous well logging work experience

ANNUAL SAFETY REVIEW (REFRESHER TRAINING) [§39.61(c) and (d)]

[ ] Description of topics covered

[ ] Instructor name

ANNUAL JOB PERFORMANCE AUDIT OF WELL LOGGING SUPERVISORS [§39.13(d)]

[ ] Description of the program

[ ] Discussion of management action

[ ] Commitment to inspect each logging supervisor at intervals not to exceed 1 year

[ ] Inspections made on-the-job & unannounced

[ ] Commitment that an individual who has not performed logging for more than 1 year will be inspected the first time that person engages in logging operations

[ ] Name, training, & experience of each person who will conduct inspection

ITEM 8.9   **FACILITIES AND EQUIPMENT**

Facility: For Each Field Station

[ ]   Sketch/drawing *to scale* of the facility and all work areas where materials (tracer or sealed source) will be used or stored

[ ]   Identify the following, where applicable:

    ( )   Areas where explosive, flammable or other hazardous materials stored

    ( )   Buildings

    ( )   Boundary lines

    ( )   Security fences

    ( )   Local storage areas

    ( )   Drawn to specified scale

    ( )   Sketch/drawing of:

        < >   locked storage container

        < >   underground storage bunker

        < >   security of licensed materials

Facility: For Tracer Authorization, Provide

[ ]   Ready-to-use form

[ ]   Bench top preparation

    ( )   Describe laboratory areas for sample preparation

        < >   Hoods

        < >   Hood filters

        < >   Sinks

        < >   Trays with absorbent materials

        < >   Remote handling tools

        < >   Rubber gloves

[ ]   Storage provisions

    ( )   Describe & provide a drawing of storage facilities

    ( )   Storage of waste materials included

    ( )   Security provisions

    ( )   Adequate shielding

[ ]   General safety equipment available at temporary job sites: [§39.45(a)]

    ( )   rubber gloves

    ( )   face shield

    ( )   respirator

    ( )   coveralls

      ( ) auxiliary shielding

      ( ) absorbent material

      ( ) secondary container

      ( ) plastic bag

[ ] Laundry Facility for contaminated clothing, etc.

[ ] Decontamination Facilities for trucks, tracer injection tools, or other equipment contaminated by tracer materials.

ITEM 8.10.1   AGREEMENT WITH WELL OWNER/OPERATOR [§39.15]

[ ] Elements of the Agreement:

    ( ) A reasonable effort be made to recover the source

    ( ) A person not attempt to recover a lodged sealed source in a manner which, in the licensee's opinion, could result in its rupture

    ( ) Radiation monitoring be conducted during recovery operations

    ( ) Contaminated equipment, personnel, or environment be decontaminated

    ( ) Irretrievable classified sources:

        < >   Means to prevent inadvertent intrusion on the source

        < >   Plaque

    ( ) Agreement refers to §39.15(a)

    ( ) Blanket agreement

    ( ) Emergency Abandonment of DTS or MWD/LWD sealed sources.

    ( ) Abandonment of Neutron Generator with activity *greater than* 110 GBq (30 curies)

    ( ) ECSs with activity *greater than* 3.7 MBq (100 microcuries)

ITEM 8.10.2   RADIATION SAFETY PROGRAM AUDIT

[ ] Reviewed on an annual basis

    ( ) ALARA

    ( ) NRC/DOT regulations & License

    ( ) Occupational/Public Doses

[ ] Audit program *not submitted*, but available for inspection by NRC

[ ] Appendix G reviewed

ITEM 8.10.3   **RADIATION MONITORING INSTRUMENTS [§39.33(a)]**

[ ]   0.1 - 50 mR/h

[ ]   Type of instruments (GM, Ion chamber, scintillation)

[ ]   Type of radiation detected (α,β,γ,neutron)

[ ]   Availability of survey instrument pursuant to 10 CFR 39.33(b)

Survey Instrument
 	
• **Manufacturer**
• **Model No.**
• **# Available**
• **Type**
  – **GMI on-chamber**
  – **Scintillation**
	Instrument Probes
	
	• **Model No.**
		Range
		
		• **CPM**
		• **DPM**
		• **mR/hr**
		• **mr/hr**
			Radiation Detected
			
			• α
			• β
			• γ
			• neutron

Counting Equipment For: • **Analysis of Contamination Swipes** • **Analysis of Bioassay Samples**		Calibration Standards	Minimum Detectable Activity
Special Equipment • **Air Samplers** • **Direct Reading Dosimeters** • **Condenser R meter**		# Available	Description

CALIBRATION OF RADIATION DETECTION INSTRUMENTS [§39.33(c)]

[ ]   6-month calibration frequency

[ ]   In-house

[ ]   By manufacturer

[ ]   By outside firm

Name _____

License No. _____

[ ]   Calibration procedures in Appendix N adopted

[ ]   Alternative calibration procedures for radiation detection instruments provided for NRC review

ITEM 8.10.4　　**MATERIAL RECEIPT AND ACCOUNTABILITY/PHYSICAL INVENTORY [§39.37]**

　　　[ ]　Semiannual frequency

　　　[ ]　Maintain records of receipt, transfer, and disposal

　　　[ ]　Required Information

　　　　　( )　Quantity and kind of licensed material (Sealed Source/Tracer)

　　　　　( )　Location of licensed material

　　　　　( )　Date of inventory

　　　　　( )　Name of individual conducting inventory

　　　　　( )　Inventory records for sealed sources may be combined with leak test records

ITEM 8.10.5　　**PERSONNEL MONITORING [§39.65(a)]**

　　　　( )　TLD

　　　　( )　Film

　　　　( )　OSL - *Note: Exemption should be requested*

　　　　( )　Neutron capability

　　　　( )　NVLAP-Approved

　　　　( )　Exchange frequency

　　　　　　< >　Monthly

　　　　　　< >　Quarterly

BIOASSAYS [§39.65(b)]

　　　[ ]　Procedures in RG 8.20 adopted for conducting bioassays

　　　[ ]　Commitment not to expose any individual to 50 mCi of I-131 at a time or in any 5 days

　　　[ ]　Commercial Service:

　　　　　( )　Name _____

　　　　　( )　License No. _____

ITEM 8.10.6　　**PUBLIC DOSE**

　　　[ ]　No response required

　　　[ ]　Appendix P reviewed

ITEM 8.10.7   **OPERATING AND EMERGENCY PROCEDURES [§39.63]**

[ ] Instructions for handling and using licensed materials, including sealed sources in wells, without surface casing for protecting fresh water aquifers

[ ] Instructions for maintaining security during storage and transportation

[ ] Instructions to keep licensed material under control and immediate surveillance during use

[ ] Steps to take to keep radiation exposures ALARA

[ ] Steps to maintain accountability during use

[ ] Steps to control access to work sites

[ ] Steps to take and whom to contact when an emergency occurs

[ ] Instructions for using remote handling tools when installing into well logging tools or handling sealed sources when returning them to their transport containers. Although good information, instructions are not necessary when handling low-activity calibration sources and radioactive tracer materials.

[ ] Methods and occasions for conducting radiation surveys, including surveys for detecting Contamination, as required by 10 CFR 39.67(c) - (e)

[ ] Procedures to minimize personnel exposure during routine use and in the event of an incident, including exposures from inhalation and ingestion of licensed tracer materials

[ ] Methods and occasions for locking and securing stored licensed materials

[ ] Personnel monitoring, including bioassays, and the use of personnel monitoring equipment

[ ] Transportation of licensed materials to field stations or temporary job sites, packaging of licensed materials for transport in vehicles, placarding of vehicles when needed, and physically securing licensed materials in transport vehicles during transportation to prevent accidental loss, tampering, or unauthorized removal

[ ] Procedures for picking up, receiving, and opening packages containing licensed materials, in accordance with 10 CFR 20.1906

[ ] Instructions for the use of tracer materials, how to decontaminate the environment, equipment, and personnel

[ ] Instructions for maintaining records in accordance with the regulations and the license conditions

[ ] Steps for the use, inspection, and maintenance of sealed sources, source holders, logging tools, injection tools, source handling tools, storage containers, transport containers, and uranium sinker bars, as required by 10 CFR 39.43

[ ] Procedures for identifying and reporting to NRC defects and noncompliance, as required by 10 CFR 21.21(a)

[ ] Actions to be taken if a sealed source is lodged in a well

[ ] Procedures and actions to be taken if a sealed source is ruptured, including actions to prevent the spread of contamination and minimize inhalation and ingestion of licensed materials and actions to obtain suitable radiation survey instruments, as required by 10 CFR 39.33(b)

[ ] Instructions for the proper storage and disposal of radioactive waste

[ ] Procedures for laundering contaminated clothing and for decontaminating equipment and vehicles

[ ] Procedures to be followed in the event of uncontrolled release of radioactive tracer material to the environment, including notification of the RSO, NRC, and other Federal and State Agencies

ITEM 8.10.8   **LEAK TESTING [§39.35]**

[ ] Vendor(s) Name: _____
Address: _____
Agreement State/NRC License No.: _____

[ ] Leak test kit

[ ] Leak testing conducted in-house using Appendix R procedures

[ ] Alternative leak testing procedures submitted for NRC review

ITEM 8.10.9   **MAINTENANCE**

[ ] *Daily* visual inspection and *6-month routine maintenance* [§39.43(a)-(b)]

( ) Source holders

( ) logging tools

( ) injection tools

( ) source handling tools

( ) storage containers

( ) transport containers

( ) uranium sinker bars

[ ] Daily [§39.43(a)]

( ) Defects (§39.43(a))

( ) Repairs made and recorded, or equipment taken out of service

( ) Operation performed by logging supervisor

SEMIANNUAL MAINTENANCE [§39.43(b)]

[ ] 6-month

( ) Defects (§39.43(b))

( ) Repairs made and recorded, or equipment taken out of service

( ) Operation performed by logging supervisor

REMOVAL OR MAINTENANCE ON A SEALED SOURCE OR HOLDER [§39.43(c)]

[ ] Services performed by manufacturer

[ ] Performed by individual licensed by Agreement State/NRC

[ ] Performed by licensee

    ( ) Detailed procedures for each task provided for NRC review

    ( ) Radiation safety precautions outlined in O&E procedures

SEALED SOURCES STUCK IN A SOURCE HOLDER [§39.43(d)]

[ ] Performed by licensed equipment manufacturer

[ ] Performed by individual licensed by Agreement State/NRC

[ ] Performed by licensee

    ( ) Detailed procedures for each task provided for NRC review

    ( ) Radiation safety precautions outlined in O&E procedures

OPENING, REPAIR, OR MODIFICATION OF SEALED SOURCES [§39.43(e)]

[ ] Performed by Agreement State/NRC-licensed firm

[ ] Performed by licensee

    ( ) Detailed procedures for each task provided for NRC review

    ( ) Radiation safety precautions outlined in O&E procedures

ITEM 8.10.10   **TRANSPORTATION**

[ ] No response required; DOT regulations will be followed

[ ] Appendix S reviewed

ITEM 8.10.11   **MINIMIZATION OF CONTAMINATION [§39.69]**

[ ] Implementation of and adherence to good health physics practices while performing operations

[ ] Minimization of distance to areas, to the extent practicable, where licensed materials are used and stored

[ ] Maximization of survey frequency, within reason, to enhance detection of contamination

[ ] Segregation of radioactive material in waste storage areas

[ ] Segregation of sealed sources and tracer materials to prevent cross-contamination

[ ] Separation of radioactive material from explosives

[ ] Separation of potentially contaminated areas from clean areas by barriers or other controls

[ ] **Request to Perform Major Decontamination Activities**

[ ] Instructions to personnel on how to determine presence through survey

[ ] Levels of contamination

[ ] Decontamination procedures

[ ] Decontamination equipment

[ ] Prevention of contamination of personnel during decontamination

[ ] How to handle contaminated waste materials

[ ] Re-survey of contaminated area to determine effectiveness

[ ] Records of survey

[ ] Before

[ ] After

[ ] Contact person

[ ] **Decontamination activities will be conducted by outside sources licensed by NRC or an Agreement State to conduct these activities.**

ITEM 8.10.12   **SEALED SOURCES**

DRILL-TO-STOP WELL LOGGING OPERATIONS

[ ] Step-by-step O&E procedures provided for NRC review

[ ] Summary or outline addressing important aspects of O&E procedures provided for review

[ ] For use of sealed sources in well without surface casing

( ) Knowledge of borehole conditions

( ) Caliper log

( ) Running dummy tool log

( ) Temporary casing

MWD/LWD WELL LOGGING OPERATIONS

[ ] Step-by-step O&E procedures provided for NRC review

[ ] Summary or outline addressing important aspects of O&E procedures provided for review

ENERGY COMPENSATION SOURCES

[ ] Step-by-step O&E procedures provided for NRC review

[ ] Summary or outline addressing important aspects of O&E procedures provided for review

    ( ) Instructions for testing ECSs requiring leak tests at intervals not to exceed 3 years

    ( ) Instructions for conducting physical inventories of ECSs at least every 6 months

    ( ) A system for maintaining inventory records required by 10 CFR 39.37

    ( ) A system for maintaining records of use for ECSs

[ ] For use of ECSs in well without surface casing

    ( ) Knowledge of borehole conditions

    ( ) Caliper log

    ( ) Running dummy tool log

    ( ) Temporary casing

ITEM 8.10.13   **TRACER STUDIES**

Tracer Studies in Single Well Applications [§39.45]

[ ] Methods and occasions for conducting radiation surveys

[ ] Methods and occasions for locking and securing tracer materials

[ ] Personnel monitoring and the use of personnel monitoring equipment

[ ] Transportation to temporary job sites and field stations, including the packaging and placing of tracer materials in vehicles, placarding of vehicles, and securing tracer materials during transportation

[ ] Procedures for minimizing exposure to members of the public and occupationally exposed individuals in the event of an accident

[ ] Maintenance of records at field stations and temporary job sites

[ ] Use, inspection, and maintenance of equipment (injector tools, remote handling tools, transportation containers, etc.)

[ ] Procedures to be used for picking up, receiving, and opening packages containing radioactive material

[ ] Decontamination of the environment, equipment, and personnel

[ ] Notifications of proper personnel in the event of an accident.

Field flood and Secondary Recovery Applications

[ ] Field flood or Secondary Recovery Applications will not be conducted

[ ] Agreement with well operator or owner, although not required by NRC regulations, is a good practice

[ ] Field flood study project design

[ ] Pre-injection phase of the field flood project

[ ] Injection phase

[ ] Post-injection phase

[ ] Emergency procedures

[ ] Reporting and record keeping requirements

[ ] Waste management

[ ] Methods and occasions for conducting radiation surveys

[ ] Methods and occasions for locking and securing tracer materials

[ ] Personnel monitoring and the use of personnel monitoring equipment

[ ] Transportation to temporary job sites and field stations, including the packaging and placing of tracer materials in vehicles, placarding of vehicles, and securing tracer materials during transportation

[ ] Procedures for minimizing exposure to members of the public and occupationally exposed individuals in the event of an accident

[ ] Maintenance of records at field stations and temporary job sites

[ ] Use, inspection and maintenance of equipment (injector tools, remote handling tools, transportation containers, etc.)

[ ] Procedures to be used for picking up, receiving, and opening packages containing radioactive material

[ ] Decontamination of the environment, equipment, and personnel

[ ] Notifications of proper personnel in the event of an accident

[ ] Information requested in Appendix F provided

Tracer Studies in Single Well Applications in Fresh Water Aquifers [§39.45]

[ ] Tracer Studies in Single Well Application will not be conducted in Fresh Water Aquifers

[ ] Tracer Studies in Single Well Application will be conducted in Fresh Water Aquifers, and an environmental report is provided for NRC's review

ITEM 8.10.14   **RADIOACTIVE COLLAR AND SUBSIDENCE OR DEPTH CONTROL MARKERS [§39.47]**

[ ] Operating and emergency procedures must include a commitment that radioactive markers can be used only where each individual marker contains quantities of licensed material not exceeding the quantities identified in 10 CFR 30.71, Schedule B

[ ] Licensees are not restricted to using only one marker, and may use multiple markers in each pipe joint, provided each individual marker (wires, tape, nails, etc.) is *not greater than the quantities identified in 10 CFR 30.71*

[ ] Provisions included in O&E procedures to ensure that radioactive markers be physically inventoried at intervals not to exceed 6 months, as specified in 10 CFR 39.37

ITEM 8.10.15   **NEUTRON ACCELERATORS USING LICENSED MATERIAL**

[ ]   Neutron generator tubes are not considered well logging sealed sources and are not required to satisfy the requirement for well logging sealed sources

[ ]   Neutron generator tubes containing less than 110 GBq (30 curies) of tritium are:

( )   Exempt from leak testing requirements if they contain less than 3.7 MBq (100 microcuries)

( )   Exempt from abandonment requirements

( )   Exempt from the performance requirements of sealed sources used in well logging operations

( )   Neutron generators containing target sources greater than 100 GBq (30 curies) cannot be used in wells without surface casing to protect fresh water aquifers, unless approved by NRC

( )   O&E procedures address handling of contamination resulting from the routine use, initial installation, replacement, or accidental damage of the targets or glass tubes

ITEM 8.10.16   **DEPLETED URANIUM [§40.51]**

[ ]   Depleted uranium sinker bars will be obtained under the provisions of a general license per 10 CFR 40.51, and registration form NRC Form 244 will be filed, as required

[ ]   Depleted uranium sinker bars will not be obtained under the provision of general license 10 CFR 40.51

[ ]   Uranium sinker bars will be possessed and inspected as specified

[ ]   Number of kilograms of specifically licensed DU specified

ITEM 8.11   **WASTE MANAGEMENT [10 CFR Part 20, Subpart K]**

[ ]   Decay-in-storage disposal for radioactive materials with half-lives less than or equal to 120 days

( )   When a container is transferred to the waste storage area, mark the container with an identification label that includes the date sealed, the isotope in the container, and the initials of the person sealing the container

( )   <120 day T½ material

( )   Held for decay a minimum of 10 T½

( )   Confirm that prior to disposal as in-house waste, you will monitor each container, as follows:

< >   Check radiation detection survey meter for proper operation

< >   Monitor container in a low-level area (less than 0.05 mrem/hr)

< >   Remove any shielding from container

< >   Monitor all surfaces

      < >   Discard only those containers that cannot be distinguished from background

      < >   Container that can be distinguished from background must be returned to storage area for further decay or transferred to person licensed to receive such waste

[ ] Return to manufacturer

[ ] Extended Interim Storage of materials pending disposal or transfer to authorized recipient

[ ] Licensed company

[ ] Sand-out, flowback, screenout, etc.

[ ] Disposal by release into sanitary sewerage (§20.2003)

[ ] Appendix T reviewed

ITEM 8.12   FEES

[ ] Fee, if any required, attached

ITEM 8.13   CERTIFICATION

[ ] Individual signing an application authorized to make binding commitments and to sign official documents on behalf of the legal entity or applicant

Appendix E

Sample License

NRC FORM 374

U.S. NUCLEAR REGULATORY COMMISSION

MATERIALS LICENSE

Pursuant to the Atomic Energy Act of 1954, as amended, the Energy Reorganization Act of 1974 (Public Law 93-438), and Title 10, Code of Federal Regulations, Chapter I, Parts 30, 31, 32, 33, 34, 35, 36, 39, 40, and 70, and in reliance on statements and representations heretofore made by the licensee, a license is hereby issued authorizing the licensee to receive, acquire, possess, and transfer byproduct, source, and special nuclear material designated below; to use such material for the purpose(s) and at the place(s) designated below; to deliver or transfer such material to persons authorized to receive it in accordance with the regulations of the applicable Part(s). This license shall be deemed to contain the conditions specified in Section 183 of the Atomic Energy Act of 1954, as amended, and is subject to all applicable rules, regulations, and orders of the Nuclear Regulatory Commission now or hereafter in effect and to any conditions specified below.

Licensee

1

2

3. License number

4. Expiration date

5. Docket No.

Reference No.

6. Byproduct, source, and/or special nuclear material	7. Chemical and/or physical form	8. Maximum amount that licensee may possess at any one time under this license
A. Americium-241	A. Sealed neutron sources [insert manufacturer and model numbers]	A. Not to exceed [No]. curies per source
B. Cesium-137	B. Sealed sources [insert manufacturer and model numbers]	B. Not to exceed [No]. curies per source
C. Iodine-131	C. Liquid	C. Not to exceed [No]. millicuries total and [No]. millicuries per injection
D. Iridium-192	D. "Labeled" Frac Sands	D. Not to exceed [No]. millicuries total and [No]. millicuries per injection

9. **Authorized use:**

A. through E.   For use in oil and gas well logging.

NRC FORM 374A U.S. NUCLEAR REGULATORY COMMISSION	PAGE 2 of 3 PAGES
MATERIALS LICENSE SUPPLEMENTARY SHEET	License Number Docket or Reference Number

CONDITIONS

10. Radioactive material shall be used only at the following:

    A. [permanent site address]

    B. Temporary job sites anywhere in the United States where the U.S. Nuclear Regulatory Commission maintains jurisdiction for regulating licensed material, including areas of exclusive Federal jurisdiction within Agreement States.

    If the jurisdiction status of a Federal facility within an Agreement State is unknown, the licensee should contact the federal agency controlling the job site in question to determine whether the proposed job site is an area of exclusive Federal jurisdiction. Authorization for use of radioactive materials at job sites in Agreement States not under exclusive Federal jurisdiction shall be obtained from the appropriate state regulatory agency.

11. The licensee shall not vacate or release to unrestricted use a field office or storage location whose address is identified in Condition 10, without prior U.S. Nuclear Regulatory Commission approval.

    A. Licensed material shall be used by, or under the supervision and in the physical presence of, or individuals who have been trained as specified in [( ) letter, ( ) application, ( ) facsimile] [( ) dated, ( ) received] [TYPE IN DATE OF APPLICATION OR LETTER].

    B. The Radiation Safety Officer for this license is [TYPE IN NAME OF INDIVIDUAL].

12. Notwithstanding the periodic leak test required by 10 CFR 39.35, the requirement does not apply to sources, except sources containing plutonium, that are stored and not being used. The sources exempted from this periodic test shall be tested for leakage before use or transfer to another person. No sealed source shall be stored for a period of more than 10 years without being tested for leakage and/or contamination.

13. Each source holder or logging tool containing radioactive material shall bear a legible and visible marking as specified in 10 CFR 39.31(a). The label must be on the smallest component that contains the licensed material which is transported as a separate piece of equipment.

14. The opening, repair, or modification of any Energy Compensation Source must be performed by persons specifically approved to do so by the Commission or an Agreement State.

NRC FORM 374A	U.S. NUCLEAR REGULATORY COMMISSION	PAGE 3 of 3 PAGES

MATERIALS LICENSE
SUPPLEMENTARY SHEET

License Number

Docket or Reference Number

15. In addition to the possession limits in Item 8, the licensee shall further restrict the possession of licensed material to quantities below the minimum limit specified in 10 CFR 30.35(d) for establishing decommissioning financial assurance.

16. Except as specifically provided otherwise in this license, the licensee shall conduct its program in accordance with the statements, representations, and procedures contained in the documents, including any enclosures, listed below. The Nuclear Regulatory Commission's regulations shall govern unless the statements, representations, and procedures in the licensee's application and correspondence are more restrictive than the regulations.+-

A.

FOR THE U.S. NUCLEAR REGULATORY COMMISSION

Date _____   By [Insert Reviewer's Name and Title]

Nuclear Materials Licensing Branch
Region IV
Arlington, Texas 76011

Appendix F

Field Flood Studies/Enhanced Recovery of Oil and Gas Wells

Field Flood Studies/Enhanced Recovery of Oil and Gas Wells

A formal contractual agreement with well operator or owner should specify control points at which samples will be taken, establish criteria for setting minimum sample requirements, and confirm the willingness of the client company to abide by effluent restrictions and undertake remedial action, if required. Following are some examples:

Samples of recovered fluids or gas will be collected and measured according to the established sampling schedule.

- Appropriate remedial action will be taken if accidents or incidents occurred that may result in the release of licensed materials to the environment. For example, if the concentration in the recovered fluid or gas approaches or exceeds the design limits, remedial action should be taken, such as reducing the injection pressure, temporarily shutting in the well, or diluting with non-tracer-bearing gas.

Planning Stage

Reservoir Information

Describe the reservoir information that you need in order to design a radioisotope tracer study for a field flood operation. Examples of reservoir information are shown below:

- Reservoir volume
- Reservoir thickness
- Porosity
- Injected volumes (liquids/gases)
- Oil/water saturation ratios

Project Design

Outline the design of the tracer application requested. Examples of items to consider are the following:

- Choice of radionuclides and method used to determine (1) the amount of radionuclide to be injected, and (2) the expected concentration of radionuclide in the fluids (gas, water, oil) at a recovery well site. Indicate your adherence to the ALARA principle
- How breakthrough time is predicted

- How tracer concentrations in the recovered liquids and gases are estimated

- How the sampling schedule at production wellheads is determined. Include a description of how you would determine when sampling could be discontinued. As an example, monitoring of samples may be ended when the design life of the project is completed, unless the effluent concentration at the control point is above a specified fraction of the maximum permissible concentration (as listed for unrestricted areas in 10 CFR 20) and is increasing. In that case, the control point will be monitored until the concentration is below the specified fraction of the annual average concentration specified in 10 CFR Part 20, Appendix B, Table 2.

Pre-injection Stage

Transportation of licensed materials. State that the applicant will comply with NRC and DOT regulations pertaining to the transportation of licensed material. Particular attention should be directed to monitoring requirements upon receipt of packages containing licensed materials.

Integrity of wellhead assembly and wellbore. Describe the test procedures used to ensure that the wellhead assembly, including injection equipment, will not leak under operating conditions. Describe the procedures used to ensure that the wellbore will not leak underground. For example, if the injection well operates properly for a 2-week period, integrity of the wellbore may be considered ensured.

Injection Stage

Outline radiation safety practices during injection process.

Following are examples of practices:

- Remain upwind, if practical.

- Keep nonessential personnel at a distance.

- Use personnel monitoring devices (TLD, OSL, film badges, finger badges, pocket dosimeters, etc.) and other radiation detection instruments in your monitoring and surveillance programs.

- Use special tools and devices to handle licensed material and to facilitate the injection process.

- Perform visual inspection, check pressure gauge, etc., to assure absence of leaks and proper delivery of injection liquid or gas.

- Continuously or intermittently monitor radiation levels outside the injection assembly to assure that the injection is proceeding according to the plan. Allow sufficient time before opening wellhead assembly.

Post-injection Stage

Outline radiation safety practices that will be put into place after the injection phase is completed. Examples of practices include the following:

- Check exposure rate at wellhead assembly for residual activity.

- Take smear samples to detect removable contamination on wellhead assembly.

- Clean reusable tools and check for residual activity before securing for reuse.

- Collect contaminated materials or contaminated tools and package them into an appropriate waste container.

- Establish schedule for taking samples for bioassay when, for example, handling tritium (H-3) exceeding 3.7 Gbq (0.1 Ci) or gaseous H-3 exceeding 3,700 Gbq (100 Ci), or handling radioiodine exceeding 1.85 Gbq (50 mCi) of iodine-131 or iodine-125.

- Provide instructions to well operator's personnel for taking post-injection samples and shipping the samples to your facilities for analysis. Include handling, packaging, and shipping procedures.

- Package waste materials for transportation, prepare appropriate labels and shipping papers, and check for radiation level and removable contamination outside the package.

- Measure concentrations of radionuclides in recovered liquids or gases, according to your established sampling schedule.

- Take corrective measures if the concentrations in the recovered liquids or gas approach or exceed design levels.

- Conduct area and personnel monitoring before leaving injection site.

Emergency Procedures

Outline procedures that you will follow in the event of incidents or accidents that release radioactive materials to the environment. Following are examples of incidents and accidents:

- Discovering a leaky source in a shipping container

- Dropping and breaking a source container, thereby spilling the source on the ground

- Detecting leakage of radioactive materials from wellhead assembly

- Measuring concentrations in liquids or gas from production wells exceeding limits specified in Table 2, Appendix B, 10 CFR Part 20.

Reporting, Record Keeping, and Notification

Outline the report that will be submitted to the NRC and the records maintained regarding the field flood injections. Following are examples of releases to include: records on the identification of wells, radionuclides, and quantities injected; concentrations of radionuclides in liquids or gases produced at production wells; and concentrations of radionuclides in products released from the field.

Also outline the procedures you will follow in case of accidents; and procedures for notifying the proper persons or organizations, such as your company management (RSO), well operator or owner, and Federal, State, or municipal Governmental Agencies involved with the control and oversight of affected wells.

Waste Management

The applicant should outline the procedures for disposing of licensed material. Wastes from tracer operations such as unused materials, and contaminated wipes, gloves, tools, clothing, containers, etc., should be disposed of in accordance with 10 CFR Part 20. Recovered waste fluids that contain radioactive tracers should either be reinjected or treated as radioactive waste. A commonly used method of disposal is transfer to a commercial firm licensed by NRC or an Agreement State to accept radioactive wastes. In dealing with these firms, prior contact is needed to determine the specific services they can provide. If commercial services will be used, this should be specified.

Reporting, Record Keeping, and Notification

Outline the report that will be submitted to the NRC and the records maintained regarding the field flood injections. Following are examples of releases to include: records on the identification of wells, radionuclides, and quantities injected; concentrations of radionuclides in liquids or gases produced at production wells; and concentrations of radionuclides in products released from the field.

Also outline the procedures you will follow in case of accidents; and procedures for notifying the proper persons or organizations, such as your company management (RSO), well operator or owner, and Federal, State, or municipal Governmental Agencies involved with the control and oversight of affected wells.

Waste Management

The applicant should outline the procedures for disposing of licensed material. Wastes from tracer operations such as unused materials, and contaminated wipes, gloves, tools, clothing, containers, etc., should be disposed of in accordance with 10 CFR Part 20. Recovered waste fluids that contain radioactive tracers should either be reinjected or treated as radioactive waste. A commonly used method of disposal is transfer to a commercial firm licensed by NRC or an Agreement State to accept radioactive wastes. In dealing with these firms, prior contact is needed to determine the specific services they can provide. If commercial services will be used, this should be specified.

Appendix G

Suggested Well Logging and Field Flood Audit Checklist

Suggested Well Logging and Field Flood Audit Checklist

All areas indicated in audit notes may not be applicable to every license and may not need to be addressed during each audit. For example, licensees do not need to address areas that do not apply to the licensee's activities, and activities that have not occurred since the last audit need not be reviewed at the next audit.

Date of This Audit_____   Date of Last Audit _____

Next Audit Date _____

Auditor _____   Date _____
                    (Signature)

Management Review _____   Date _____
                            (Signature)

Type of Inspection: ( ) Announced   ( ) Unannounced

Summary of Findings and Actions

      [ ] No violations cited

      [ ] Self-identified Violation(s)

      [ ] Concerns

A. ORGANIZATION AND SCOPE OF PROGRAM
Organization and scope of radiation safety program in accordance with application and the license.

B. MANAGEMENT OVERSIGHT

1. Radiation Safety Officer

2. Audits, Reviews, or Inspections

10 CFR 20.1101	Radiation protection programs.
10 CFR 20.2102	Records of radiation protection programs.

Audits required by license conditions.

3. Use by Authorized Individuals.

   Management structure and control as specified in the license.

4. ALARA

   10 CFR 20 1101   ALARA program.

C. FACILITIES

1. Facilities as Described.

   Facilities as described in the license.

2. Storage

10 CFR 20.1801	Security of stored material.
10 CFR 39.31	Labels, security, and transportation precautions.

D. EQUIPMENT AND INSTRUMENTATION

1. Instruments and Equipment

10 CFR 39.33	Radiation detection instruments.

   Radiation detection instruments and equipment as described in the license.

2. Sources, Source Holders, Tools

10 CFR 39.31	Labels, security and transportation precautions.
10 CFR 39.49	Uranium sinker bars.

   Equipment and instrumentation as specified in the license.

E. MATERIAL USE, CONTROL, AND TRANSFER

1. Security and Control

10 CFR 20.1003	Definitions (restricted area and unrestricted area).
10 CFR 20.1801	Security of stored material.

| 10 CFR 20.1802 | Control of material not in storage. |
| 10 CFR 39.71 | Security. |

2. Receipt and Transfer of Licensed Material

10 CFR 20.1302	Compliance with dose limits for individual members of the public.
10 CFR 20.1906	Procedures for receiving and opening packages.
10 CFR 20.1501	General.
10 CFR 20.2103	Records of surveys.
10 CFR 30.41	Transfer of byproduct material.
10 CFR 30.51	Records of receipt and transfer.

3. Isotope, Chemical Form, Quantity, and Use

| 10 CFR 39.37 | Physical inventory. |
| 10 CFR 39.47 | Radioactive markers. |

Receipt and transfer as described in the license.

F. INSPECTION AND MAINTENANCE

| 10 CFR 39.43 | Inspection, maintenance, and opening of a source or source holder. |
| 10 CFR 21.21 | Notification of failure to comply or existence of a defect and its evaluation. |

Inspection and maintenance as described in the license.

G. AREA RADIATION SURVEYS AND CONTAMINATION CONTROL

1. Area Surveys

10 CFR 20.1302	Compliance with dose limits for individual members of the public.
10 CFR 20.1501	General.
10 CFR 20.2103	Records of surveys.

10 CFR 20.2107	Records of dose to individual members of the public.
10 CFR 39.67	Radiation surveys.
10 CFR 39.69	Radioactive contamination control.

Area radiation surveys and contamination control as described in the license.

2. Leak Tests and Inventories

10 CFR 39.35	Leak testing of sealed sources.

Leak test conducted in accordance with applicable license conditions.

3. Tracer Studies

10 CFR 39.45	Subsurface tracer studies.
10 CFR 39.51	Use of a sealed source in a well without a surface casing.

H. TRAINING AND INSTRUCTIONS TO WORKERS

General

10 CFR 19.12	Instruction to workers.
10 CFR 39.61	Training.

Knowledge of 10 CFR Part 20 radiation protection procedures and requirements.

Training program for personnel in accordance with the license.

I. RADIATION PROTECTION

1. Radiation Protection Program

   a. Exposure evaluation

10 CFR 20.1501	General.

   b. Programs

10 CFR 20.1101	Radiation protection programs.

2. Dosimetry

   a. Dose Limits

10 CFR 20.1201	Occupational dose limits for adults.
10 CFR 20.1202	Compliance with requirements for summation of external and internal doses.
10 CFR 20.1207	Occupational dose limits for minors.
10 CFR 20.1208	Doses to an embryo/fetus.

   b. External

10 CFR 39.65	Personnel Monitoring.
10 CFR 20.1203	Determination of external dose from airborne radioactive material.
10 CFR 20.1501	General.
10 CFR 20.1502	Conditions requiring individual monitoring of external and internal occupational dose.

   Dosimetry provided in accordance with the license.

   c. Internal

10 CFR 39.65	Personnel Monitoring
10 CFR 20.1204	Determination of internal exposure.
10 CFR 20.1502	Conditions requiring individual monitoring of external and internal occupational dose.
10 CFR Part 20, Subpart H	Respiratory protection and controls to restrict internal exposure in restricted areas.

3. Records

10 CFR 20.2102	Records of radiation protection programs.
10 CFR 20.2103	Records of surveys.
10 CFR 20.2104	Determination of prior occupational dose.
10 CFR 20.2106	Records of individual monitoring results.

J. RADIOACTIVE WASTE MANAGEMENT

1. Disposal

10 CFR 30.41	Transfer of byproduct material.
10 CFR 20.1904	Labeling containers.
10 CFR 20.2001	General requirements.
10 CFR 20.2103	Records of surveys.
10 CFR 20.2108	Records of waste disposal.
10 CFR 20.2003	Disposal by release into sanitary sewerage.

2. Effluents

   a. General

   Maintaining Effluents from Materials Facilities as Low as Is Reasonably Achievable (ALARA).

   b. Release to septic tanks

10 CFR 20.1003	Definitions (sanitary sewerage).
10 CFR Part 20,	
App B, Table 2	Effluent concentrations.

   c. Incineration of waste

10 CFR 20.2004	Treatment or disposal by incineration.

   d. Control of air effluents and ashes

10 CFR 20.1201	Occupational dose limits for adults.
10 CFR 20.1301	Dose limits for individual members of the public.
10 CFR 20.1501	General.
10 CFR 20.1701	Use of process or other engineering controls.

   Incineration conducted in accordance with license condition.

3. Waste Management

    a. General

       10 CFR 20.2001   General requirements.

       Radioactive Waste Management - Inspection of Waste Generator Requirements of 10 CFR Part 20 and 10 CFR Part 61.

    b. Waste compacted

       Applicable license conditions.

    c. Waste storage areas

       10 CFR 20.1801   Security of stored material.

       10 CFR 20.1902   Posting requirements.

       10 CFR 20.1904   Labeling containers.

       Waste storage areas in accordance with the license.

    d. Packaging, Control, and Tracking

       10 CFR Part 20, Appendix G   Requirements for Transfers of Low-Level-Waste Intended for Disposal at Land Disposal Facilities and Manifests.

       10 CFR 20.2006   Transfer for Disposal and Manifests.

       10 CFR 61.55   Waste classification.

       10 CFR 61.56   Waste characteristics.

    e. Transfer

       10 CFR Part 20, Appendix G   Requirements for Transfers of Low-Level-Waste Intended for Disposal at Land Disposal Facilities and Manifests.

       10 CFR 20.2001   General requirements.

       10 CFR 20.2006   Transfer for disposal and manifests.

f. Records

    10 CFR 20.2103   Records of surveys.

    10 CFR 20.2108   Records of waste disposal.

K. DECOMMISSIONING

10 CFR 30.36   Expiration and termination of licenses and decommissioning of sites and separate buildings or outdoor areas.

L. TRANSPORTATION

1. General

Hazard Communication for Class 7 (Radioactive) Materials.

10 CFR 71.5   Transportation of licensed material.

Implementation of Revised 49 CFR Parts 100-179 and 10 CFR Part 71.

2. Shippers - Requirements for Shipments and Packaging

    a. General Requirements

        49 CFR Part 173, Subpart I   Class 7 (radioactive) materials.

        49 CFR 173.24   General requirements for packaging and packages.

        49 CFR 173.448   General transportation requirements.

        49 CFR 173.435   Table of A_1 and A_2 values for radionuclides.

    b. Transport Quantities

        10 CFR 71.4   Definitions.

        i.   All quantities

            10 CFR 71.4   Definitions.

            49 CFR 173.410   General design requirements.

            49 CFR 173.441   Radiation level limitations.

            49 CFR 173.443   Contamination control.

49 CFR 173.475	Quality control requirements prior to each shipment of of Class 7 (radioactive) materials.
49 CFR 173.476	Approval of special form Class 7 (radioactive) materials.

ii. Limited quantities

49 CFR 173.421	Excepted packages for limited quantities of Class 7 (radioactive) materials.
49 CFR 173.422	Additional requirements for excepted packages containing Class 7 (radioactive) materials.

iii. Type A quantities

49 CFR 173.412	Additional design requirements for Type A packages.
49 CFR 173.415	Authorized Type A packages.
49 CFR 178.350	Specification 7A; general packaging, Type A.

iv. Type B quantities

v. LSA material and SCO

49 CFR 173.403	Definitions.
49 CFR 173.427	Transport requirements for low specific activity (LSA) Class 7 (radioactive) materials and surface contaminated objects (SCO).

c. HAZMAT Communication Requirements

49 CFR 172.200-205	Shipping papers.
49 CFR 172.300-338	Marking.
49 CFR 172.400-450	Labeling.
49 CFR 172.500-560	Placarding.
49 CFR 172.600-604	Emergency response information.

3. HAZMAT Training

49 CFR 172.702	Applicability and responsibility for training and testing.
49 CFR 172.704	Training requirements.

4. Transportation by Public Highway

49 CFR 171.15	Immediate notice of certain hazardous materials incidents.
49 CFR 171.16	Detailed hazardous materials incident reports.
49 CFR 177.800	Purpose and scope of this part and responsibility for compliance and training.
49 CFR 177.816	Driver training.
49 CFR 177.842	Loading and unloading: Class 7 (radioactive) material.

M. NOTIFICATIONS AND REPORTS

10 CFR 19.13	Notifications and reports to individuals.
10 CFR 20.2201	Reports of theft or loss of licensed material.
10 CFR 20.2202	Notification of incidents.
10 CFR 20.2203	Reports of exposures, radiation levels, and concentrations of radioactive material exceeding the constraints or limits.
10 CFR 30.50	Reporting requirements.

N. POSTING AND LABELING

10 CFR 19.11	Posting of notices to workers.
10 CFR 21.6	Posting requirements.
10 CFR 20.1902	Posting requirements.
10 CFR 20.1903	Exemptions to posting requirements.
10 CFR 20.1904	Labeling containers.
10 CFR 20.1905	Exemptions to labeling requirements.

O. FIELD STATIONS AND TEMPORARY JOB SITES

1. Documents and Records at Field Stations

   10 CFR 39.73   Documents and records required at field stations.

   Records at field stations as required by license conditions.

2. 10 CFR 39.75   Documents and records required at temporary job sites.

   Records at temporary job sites as required by license conditions.

P. ABANDONMENT OF SOURCES

10 CFR 39.15   Agreement with well owner or operator.

10 CFR 39.77   Notification of incidents and lost sources; abandonment procedures for irretrievable sources.

Q. INDEPENDENT AND CONFIRMATORY MEASUREMENTS

R. PERSONNEL CONTACTED

No references.

NAME	TITLE	DATE OF CONTACT

Appendix H

Information Needed for Transfer of Control Application

Information Needed for Transfer of Control Application

Licensees must provide full information and obtain NRC's *prior written consent* before transferring control of the license; some licensees refer to this as "transferring the license." Provide the following information concerning changes of control by the applicant (transferor and/or transferee, as appropriate). If any items are not applicable, so state.

1. The new name of the licensed organization. If there is no change, the licensee should so state.

2. The new licensee contact and telephone number(s) to facilitate communications.

3. Any changes in personnel having control over licensed activities (e.g., officers of a corporation) and any changes in personnel named in the license such as Radiation Safety Officer, authorized users, or any other persons identified in previous license applications as responsible for radiation safety or use of licensed material. The licensee should include information concerning the qualifications, training, and responsibilities of new individuals.

4. An indication of whether the transferor will remain in non-licensed business without the license.

5. A complete, clear description of the transaction, including any transfer of stocks or assets, mergers, etc., so that legal counsel is able, when necessary, to differentiate between name changes and transfer of control.

6. A complete description of any planned changes in organization, location, facility, equipment, or procedures (i.e., changes in operating or emergency procedures).

7. A detailed description of any changes in the use, possession, location, or storage of the licensed materials.

8. Any changes in organization, location, facilities, equipment, procedures, or personnel that would require a license amendment even without the transfer of control.

9. An indication of whether all surveillance items and records (e.g., calibrations, leak tests, surveys, inventories, and accountability requirements) will be current at the time of transfer. Provide a description of the status of all surveillance requirements and records.

10. Confirmation that all records concerning the safe and effective decommissioning of the facility, pursuant to 10 CFR 30.35(g), 40.36(f), 70.25(g), and 72.30(d); public dose; and waste disposal by release to sewers, incineration, radioactive material spills, and on-site burials, have been transferred to the new licensee, if licensed activities will continue at the same location, or to the NRC for license terminations.

11. A description of the status of the facility. Specifically, the presence or absence of contamination should be documented. If contamination is present, will decontamination occur before transfer? If not, does the successor company agree to assume full liability for the decontamination of the facility or site?

12. A description of any decontamination plans, including financial assurance arrangements of the transferee, as specified in 10 CFR 30.35, 40.36, and 70.25. Include information about how the transferee and transferor propose to divide the transferor's assets, and responsibility for any cleanup needed at the time of transfer.

13. Confirmation that the transferee agrees to abide by all commitments and representations previously made to NRC by the transferor. These include, but are not limited to: maintaining decommissioning records required by 10 CFR 30.35(g); implementing decontamination activities and decommissioning of the site; and completing corrective actions for open inspection items and enforcement actions.

    With regard to contamination of facilities and equipment, the transferee should confirm, in writing, that it accepts full liability for the site, and should provide evidence of adequate resources to fund decommissioning; or the transferor should provide a commitment to decontaminate the facility before transferring control.

    With regard to open inspection items, etc., the transferee should confirm, in writing, that it accepts full responsibility for open inspection items and/or any resulting enforcement actions; or the transferee proposes alternative measures for meeting the requirements; or the transferor provides a commitment to close out all such actions with NRC before license transfer.

14. Documentation that the transferor and transferee agree to the transfer of control of the licensed material and activity, and the conditions of transfer; and that the transferee is made aware of all open inspection items and its responsibility for possible resulting enforcement actions.

15. A commitment by the transferee to abide by all constraints, conditions, requirements, representations, and commitments identified in the existing license. If not, the transferee must provide a description of its program, to ensure compliance with the license and regulations.

Appendix I

Guidance on Decommissioning Funding Plan and Financial Assurance

Guidance on Decommissioning Funding Plan and Financial Assurance

Determining Need for a Decommissioning Funding Plan and Financial Assurance

Table I.1 and the worksheet in Table I.2 are used to determine the need for certification of financial assurance (F/A) for decommissioning or a decommissioning funding plan (DFP), as required by 10 CFR 30.35. Table I.1 is a listing of isotopes with a half-life of greater than or equal to 120 days used in well logging and tracer operations. If the applicant proposes to use isotopes with a half-life greater than or equal to 120 days, divide the requested possession limit (in millicuries for unsealed material and curies for sealed sources)[1] of the isotope by the value for that isotope in Table I.1. If the material requested is in an unsealed form, use the value in the unsealed column. If the material requested is in a sealed form, use the value in the sealed column. Place the fraction in the proper column in worksheet I.2. Add the fractions in the column and place the total in the row labeled total (i.e., "sum of the ratios").

Table I.1   Isotopes With Half-lives Greater Than or Equal to 120 Days

Isotope	Quantity in Millicuries Requiring $150,000 Financial Assurance	Quantity in Millicuries Requiring $750,000 Financial Assurance	Quantity in Curies Requiring That a Decommissioning Funding Plan Be Submitted
Unsealed Licensed Material			
Calcium-45	10	100	1000
Carbon-14	100	1000	10000
Hydrogen-3	1000	10000	100000
Krypton-85	100	1000	10000
Nickel-63	10	100	1000
Silver-110m	1	10	100
Any alpha-emitting radionuclide not listed above with a half-life greater than or equal to 120 days.			

---

[1]   1 Curie = 37 gigabecquerels

Isotope	Quantity in Millicuries Requiring $150,000 Financial Assurance	Quantity in Millicuries Requiring $750,000 Financial Assurance	Quantity in Curies Requiring That a Decommissioning Funding Plan Be Submitted
Any radionuclide other than alpha-emitting radionuclide, not listed above with a half-life greater than or equal to 120 days.			
Sealed Sources			

Isotope		Quantity in Curies Requiring $75,000 of Financial Assurance
Americium-241		100
Cesium-137		100000
Cobalt-60		10000
Hydrogen-3		10000000

Table I.2   Sample Worksheet for Determining Need for a Decommissioning Funding Plan or Financial Assurance

Isotope	Unsealed Byproduct Material Activity (Millicuries) ÷ Unsealed Value from Table I.1	Sealed Byproduct Material Activity (Curies) ÷ Sealed Value from Table I.1
Total		
Funds required		

Isotope	Unsealed Byproduct Material Activity (Millicuries) ÷ Unsealed Value from Table I.1	Sealed Byproduct Material Activity (Curies) ÷ Sealed Value from Table I.1
	If < 1.0, enter $0 If > 1.0 but < 10.0, enter first level of financial assurance specified in 10 CFR 30.35(d), currently $150,000 If > 10.0, but < 100.0, enter second level of financial assurance specified in 10 CFR 30.35(d), currently $750,000 If > 100.0, enter "DFP only"	If < 1.0, enter $0 If > 1.0, enter sealed source financial assurance specified in 10 CFR 30.35(d), currently $75,000

If the sum of the fractions is less than 1 for each category (unsealed and sealed), the applicant does not need to submit certification of F/A or a DFP. If the sum of the fractions is greater than 1 for either category (sealed or unsealed), but less than 100, the applicant will need to submit certification of F/A (in the level I amount specified in 10 CFR 30.35(d), currently $150,000 or in the level II amount specified in 10 CFR 30.35(d), currently $750,000) or a DFP. If the sum of the fractions is greater than 100 for unsealed material, the applicant must submit a DFP.

Criteria Relating to Use of Financial Tests and Parent Company Guarantees for Providing Reasonable Assurance of Funds for Decommissioning" can be found in 10 CFR 30, Appendix A. "Criteria Relating to Use of Financial Tests and Self Guarantees for Providing Reasonable Assurance of Funds for Decommissioning" can be found in 10 CFR 30, Appendix C. Regulatory Guide 3.66, "Standard Format and Content of Financial Assurance Mechanisms Required for Decommissioning Under 10 CFR Parts 30, 40, 70, and 72," dated June 1990, provides sample documents for financial mechanisms. This document is currently under revision by the NRC staff.

Reference: See the Notice of Availability (on inside front cover of this report) to obtain copies of Regulatory Guide 3.66, "Standard Format and Content of Financial Assurance Mechanisms Required for Decommissioning Under 10 CFR Parts 30, 40, 70, and 72," dated June 1990.

Appendix J

NRC Letter Dated August 10, 1989, Transmitting Temporary Generic Exemptions to Well Logging Licensees

NRC Letter Dated August 10, 1989, Transmitting Temporary Generic Exemptions to Well Logging Licensees

UNITED STATES
NUCLEAR REGULATORY COMMISSION
WASHINGTON, D.C. 20555

AUG 10 1989

TO: Well Logging Licensees

FROM: John E. Glenn, Chief Medical, Academic, and Commercial Use Safety Branch
           Division of Industrial and Medical Nuclear Safety, NMSS

SUBJECT: 10 CFR PART 39.41(A)(3) TEMPORARY GENERIC EXEMPTION

Attached (Enclosure 1) is a notice of generic exemption that exempts Nuclear Regulatory Commission (NRC) well logging licensees from the requirement to use only sealed sources that meet the prototype testing requirements specified in paragraph 39.41(a)(3) of 10 CFR Part 39 in well logging operations. The exemption applies only to sealed sources that meet certain alternate prototype testing criteria.

Section 39.41 of 10 CFR Part 39 prohibits licensees from using, after July 14, 1989, a sealed source in well logging unless the source is doubly encapsulated; contains licensed material whose chemical and physical forms are as insoluble and non-dispersible as practical; and is prototype performance tested and found to maintain its integrity after each of the following tests: temperature, impact, vibration, puncture, and pressure. These prototype performance tests are the same as the tests specified for well logging sources in American National Standard Institute (ANSI) N542-1977, "Sealed Radioactive Sources, Classification," published by the National Bureau of Standards (NBS Handbook 126) in 1978. The notice also provides that NRC intends, through rulemaking, to reevaluate the requirements in Section 39.41(a)(3) for prototype testing of sealed sources. The generic exemption will allow continued use of sealed sources that were prototype tested in accordance with an earlier national standard [United States of America Standards Institute (USASI) N5.10-1968] while NRC reevaluates these requirements.

Also attached are three enclosures that list various sealed source models common to well logging and identifies their suitability for continued use in well logging operations. Enclosure 2 lists those source models which appear to meet Section 39.41 requirements and are approved for continued use. Enclosure 3 identifies those source models whose continued use is authorized under the temporary generic exemption. Enclosure 4 lists those source models that do not meet the requirements of Section 39.41 or the generic exemption and whose use in well logging must be discontinued upon receipt of this letter. When a sealed source is contained (and normally stored) within a device (logging tool), the sealed source manufacturer and model number is shown below the entry. When NRC has been able to determine that a sealed source model was manufactured/distributed by another company, or more than one model designation may have been used, this information is shown in parentheses below the entry. Neutron generators are shown by the designation "Nu GEN." An asterisk(*) indicates that the source is used within the logging tool's electronics package.

These lists may not be all inclusive; therefore, if you are authorized to use a sealed source model that is not identified on one of the lists, you should contact the individual noted below so that NRC can determine the status of the source. Upon receipt of this letter, the use of any source not listed on either Enclosure 2 or 3 must be discontinued until its suitability for continued use is determined.

Because many manufacturers are located in Agreement States, NRC relied on information from its Sealed Source and Device Registry to determine a source model '5 suitability for continued use. The Registry only summarizes the more detailed information the manufacture/distributor provides to NRC or an Agreement State when registering its sources. If you have information that shows that a source model listed on Enclosure 4 meets the requirements of Section 39.41 or the generic exemption, you may provide this information to NRC and request that the source's status be reconsidered. Alternatively, NRC will reconsider a source's status if such sources are tested and certified by a qualified testing organization as meeting Section 34.91, 10 CFR Part 39 criteria.

If you have any questions about Section 39.41, 10 CFR Part 39 regulatory requirements, the generic exemption, or the suitability of a sealed source for continued use in well logging, you should contact Bruce Carrico at (301)492-0634.

John E. Glenn, Chief
Medical, Academic, and Commercial Use
   Safety Branch
Division of Industrial and
Medical Nuclear Safety, NMSS

Enclosures: As stated

Enclosure 2

WELL LOGGING SEALED SOURCES APPROVED
UNDER PART 39 REQUIREMENTS

MANUFACTURER	MODEL
AMERSHAM CORPORATION	AMN.CYn (n = 1 to 14)
AMERSHAM CORPORATION	AMN.CY1
AMERSHAM CORPORATION	AMN.PEn (n = 1 to 4)
AMERSHAM CORPORATION	CDC.CYn (n = 2 to 12)
AMERSHAM CORPORATION	CKC.CDn (n = 2 to 12)
AMERSHAM CORPORATION	CKC.800 SERIES
AMERSHAM CORPORATION	CVN.CDn (n = 2 to 12)
AMERSHAM CORPORATION (GAMMA INDUSTRIES, GENERAL NUCLEAR)	VD (HP)
ANADRILL, INC* ISOTOPE PRODUCTS MODEL 174 SEALED SOURCE	SGS-AA, SGS-BA, OR SGS-CA
COMPROBE, INC. GAMMA INDUSTRIES MODEL VD-HP SEALED SOURCE GULF NUCLEAR, INC. MODEL VL-1 SEALED SOURCE	1203 DENSITY PROBE
DRESSER INDUSTRIES INC. (Nu GEN)	C-58301, C-1O7298
E.I.DUPONT DE NUMOURS & CO. (NEW ENGLAND NUCLEAR)	NER-571
GEARHART INDUSTRIES, INC. (Nu GEN)	013-1004-000
GENERAL ELECTRIC. CO.	GE(N)-Cf-100 SERIES
GULF NUCLEAR, INC. (NEEI)	VL-1
GULF NUCLEAR, INC. (NEEI)	71-1 (NEEI-AMBE-71-1)
KAMAN SCIENCES CORPORATION (Nu GEN)	A-3061
KAMAN SCIENCES CORPORATION (Nu GEN)	A-320

MANUFACTURER	MODEL
KAMAN SCIENCES CORPORATION (Nu GEN)	A-520
KAMAN SCIENCES CORPORATION (Nu GEN)	E-3010 AND E-3020
MONSANTO CO., DAYTON LABORATORY	H-245258 (NSR-M)
MONSANTO CO., DAYTON LABORATORY	24113
MONSANTO CO., DAYTON LABORATORY	24154-C
MONSANTO CO., DAYTON LABORATORY	24174
MONSANTO CO., DAYTON LABORATORY	24181
MONSANTO CO., DAYTON LABORATORY	24183
P.A. INCORPORATED (MONSANTO)	H-245258 (NSR-M)
P.A. INCORPORATED*	P-194693
UNC NUCLEAR INDUSTRIES	PA2A, PA2B, PT2A, PT2B, PS2A, PS2B (OLD: SM-100)
E.I. DUPONT DE NUMOURS & CO. (NEN) MODEL 478C SEALED SOURCE	
US DEPARTMENT OF ENERGY	SR-CF-100 SERIES

WELL LOGGING SEALED SOURCES APPROVED
UNDER THE GENERIC EXEMPTION

MANUFACTURER	MODEL

COMPROBE, INC.
  GULF NUCLEAR, INC. MODEL CSV SEALED
  SOURCE
	1203 DENSITY PROBE

COMPROBE, INC.
  GAMMA INDUSTRIES (GAMMATRON) MODEL
  AN-HP SEALED SOURCE
	2103 DENSITY PROBE

E.I.DUPONT DE NUMOURS & CO. (NEW ENGLAND NUCLEAR)	NER-572, NER-582
GAMMA INDUSTRIES (GENERAL NUCLEAR, INC.)	CS-1000 (HP)
GAMMA INDUSTRIES (GENERAL NUCLEAR, INC.)	GNI-NB (HP)
GAMMA INDUSTRIES	NB (HP)
GAMMA INDUSTRIES (GENERAL NUCLEAR, INC.)	NHP-A-#
GAMMA INDUSTRIES	WLG-1
GAMMATRON, INC. (NUCLEAR SOURCES AND SERVICES, INC.)	AN-HP
GAMMATRON, INC. (NUCLEAR SOURCES AND SERVICES, INC.)	AN-HPG, RN-HP
GAMMATRON, INC. (NUCLEAR SOURCES AND SERVICES, INC.)	DA-20
GAMMATRON, INC. (NUCLEAR SOURCES AND SERVICES, INC.)	DA-5
GAMMATRON, INC. (NUCLEAR SOURCES AND SERVICES, INC.)	GT-GHP
GULF NUCLEAR, INC. (NEEI)	AMBE-71-2A
GULF NUCLEAR, INC. (NEEI)	0-73-2
GULF NUCLEAR, INC. (NEEI)	CS-2

MANUFACTURER	**MODEL**
GULF NUCLEAR, INC. (NEEI)	CSV
MONSANTO CO., DAYTON LABORATORY	24112
MONSANTO CO., DAYTON LABORATORY	24120
PARKWELL LABORATORIES, INC.	PL-104

KNOWN SEALED SOURCES NOT APPROVED
FOR USE IN WELL LOGGING

MANUFACTURER	MODEL
AMERSHAM CORPORATION	CD CQ 5987
AMERSHAM CORPORATION	CDC.800 SERIES (.801 TO .811)
DRESSER ATLAS	B89596, B89587, B89598
FRONTIER TECHNOLOGY CORP.	100
GAMMA INDUSTRIES (GENERAL NUCLEAR, INC.)	GNI-DL-4
GAMMA INDUSTRIES (GENERAL NUCLEAR, INC.)	GNI-NB-S-5. 0
GAMMA INDUSTRIES	NB-S-S, NB-S-20
GAMMA INDUSTRIES (GENERAL NUCLEAR, INC.)	PL-AMBE-2.7
GAMMA INDUSTRIES	RC-1 (HP)
GAMMA INDUSTRIES	S-14
GAMMATRON, INC. (NUCLEAR SOURCES AND SERVICES, INC.)	GT-G
GENERAL NUCLEAR, INC.	GNI-C(G)M-5
GULF NUCLEAR, INC. (NEEI)	CO-50
GULF NUCLEAR, INC. (NEEI)	CS-50
GULF NUCLEAR, INC. (NEEI)	TG-1
GULF NUCLEAR, INC. (NEEI)	72-CO-200
HASTINGS RADIOCHEMICAL WORKS	CS-Ill-A-l00
ICN PHARMACEUTICAL, INC. (US NUCLEAR)	373
ICN PHARMACEUTICAL, INC. (US NUCLEAR)	374

MANUFACTURER	**MODEL**
ICN PHARMACEUTICAL, INC. (US NUCLEAR)	376
ICN PHARMACEUTICAL, INC. (US NUCLEAR)	3146
ISOTOPES SPECIALTIES	0-0037
LFE CORPORATION (TRACERLAB)	CS-15
MINNESOTA MINING AND MANUFACTURING	4F6B
MINNESOTA MINING AND MANUFACTURING	4F6H (REDESIGN OF MODEL 4F68)
MINNESOTA MINING AND MANUFACTURING	4F6S
MINNESOTA MINING AND MANUFACTURING	4P6F
MINNESOTA MINING AND MANUFACTURING	4P6U
MINNESOTA MINING AND MANUFACTURING	4P6W
MONSANTO CO., DAYTON LABORATORY (SCHLUMBERGER WELL SERVICES)	H-142525
MONSANTO CO., DAYTON LABORATORY (SCHLUMBERGER WELL SERVICES)	H-207947
MONSANTO CO., DAYTON LABORATORY	MRC
MONSANTO CO., DAYTON LABORATORY	MRC-N-SS-W-AMBE(R)
MONSANTO CO., DAYTON LABORATORY	NS-WELEX
MONSANTO CO., DAYTON LABORATORY	2410
MONSANTO CO., DAYTON LABORATORY	24154-B
NUCLEAR MATERIALS AND EQUIPMENT CORP.	NUMEC-AM-62, 63, 100, 123, 154
NUCLEAR MATERIALS AND EQUIPMENT CORP.	NUNEC DWG. 11-B-208
PARKWELL LABORATORIES, INC.	PL-AMBE
SCHLUMBERGER	DWG H-1061850
SCHLUMBERGER (MONSANTO, NUMEC)	DWG H-115686
SCHLUMBERGER	DWG H-123515
SCHLUMBERGER	DWG H-123837

MANUFACTURER	**MODEL**
SCHLUMBERGER	DWG H-142108
SCHLUMBERGER	DWG H-218733
SCHLUMBERGER	DWG H-239681
SCHLUMBERGER	DWG X-113176
SCHLUMBERGER WELL SERVICES	NSR-R
SCHLUMBERGER WELL SERVICES*	P-194693
WELL RECONNAISANCE, INC.	10411
WSI	A4794

Appendix K

Typical Duties and Responsibilities of the Radiation Safety Officer

Typical Duties and Responsibilities of the Radiation Safety Officer

The RSO's duties and responsibilities include ensuring radiological safety and compliance with NRC and DOT regulations and the conditions of the license (see Figure 8.6). Typically, these duties and responsibilities include ensuring the following. Minimum RSO duties and responsibilities include:

- Secure from management the authorization to stop activities involving licensed material considered unsafe by the RSO.

- Maintain radiation exposures ALARA.

- Develop, distribute, implement, and maintain up-to-date operating and emergency procedures.

- Ensure that the possession, installation, relocation, use, storage, repair and maintenance of licensed material and well logging equipment are consistent with the limitations in the license, the Sealed Source and Device Registration Certificate(s), and manufacturer's recommendations and instructions.

- Ensure that evaluations are performed to demonstrate that individuals who are not provided personnel monitoring devices will be unlikely to receive, in one year, a radiation dose in excess of 10% of the allowable limits or that personnel monitoring devices are provided.

- Ensure that personnel monitoring devices for well logging supervisors and assistants are used and exchanged at the proper intervals, and records of the results of such monitoring are maintained.

- Determine that licensed materials are maintained secure when not under the constant surveillance of logging personnel.

- Maintain documentation to demonstrate, by measurement or calculation, that the total effective dose equivalent to the individual likely to receive the highest dose from licensed operations does not exceed the annual limit for members of the public.

- Ensure that proper authorities are notified of incidents such as fire, theft or damage to sealed sources, loss of well logging sources down-hole, and non-routine levels of radioactive contamination at well logging, tracer, and field study operations.

- Ensure that unusual occurrences are investigated, cause(s) and appropriate corrective action(s) are identified, and timely corrective action(s) are taken.

- Perform and document radiation safety program audits annually.

- Identify violations of regulations, license conditions, or program weaknesses, and develop, implement, and document corrective actions.

- Ensure that licensed material is transported in accordance with all applicable DOT requirements.

- Ensure that licensed material is disposed of properly.

- Keep license up-to-date by amending and renewing, as required. Ensure that renewals are made in a timely manner.

- Serve as the licensee's liaison officer with the NRC on license or inspection matters.

- Control procurement and disposal of licensed material, maintain associated records, and ensure that licensed materials that are possessed or used by the applicant are limited to those specified in the license.

- Establish and conduct the training program for logging supervisors and logging assistants.

- Examine and determine the competence of logging personnel.

- Ensure that the licensed materials are used only by those individuals who have satisfactorily completed appropriate training programs or who are authorized by the license.

- Establish and maintain a personnel monitoring program and ensure that all users wear personnel monitoring equipment, such as film badges or TLD.

- Establish and maintain storage facilities.

- Establish and maintain the leak test program and supervise leak testing of sealed sources.

- Procure and maintain radiation survey instruments.

- Establish and maintain a survey instrument calibration program.

- Develop and maintain up-to-date operating and emergency procedures.

- Conduct semiannual inventories and maintain utilization logs.

- Review and ensure maintenance of those records kept by others.

- Conduct radiation safety inspections of licensed activities periodically to ensure compliance with the regulations and license conditions.

- Serve as a point of contact and give assistance in case of emergency: ( ) well logging tool damage theft ( )fire ( ) etc. to ensure that the proper authorities are notified.

- Investigate the cause of incidents and determine necessary preventative action.

- Act in an advisory capacity to the licensee's management and logging personnel.

- Establish a procedure for evaluating and reporting equipment defects and noncompliance pursuant to 10 CFR Part 21.

Appendix L

Well Logging Supervisor and Logging Assistant Training Requirements

Well Logging Supervisor and Logging Assistant Training Requirements

Table L.1   10 CFR Part 39 Training Requirements

Requirement	Training Criteria
10 CFR 39.61(a)	Logging Supervisor
A. Receive Training in 10 CFR 39.61(3) Topics (Classroom Training - Approximately 24 hours in length)	Topics in 10 CFR 39.61(e) **Fundamentals of Radiation Safety** • Characteristics of gamma radiation • Units of radiation dose and quantity of radioactivity • Hazards of exposure to radiation • Levels of radiation from licensed material • Methods of controlling radiation dose (time, distance, and shielding) • Radiation safety practices, including prevention of contamination, and methods of decontamination **Radiation Detection Instruments** • Use, operation, calibration and limitations • Survey techniques • Use of Personnel monitoring equipment **Equipment to be Used** • Operation of equipment, including source handling equipment and remote handling tools • Storage, control and disposal of licensed material • Inspection and maintenance of equipment **Requirements of Pertinent Federal Regulations** Case histories of accidents in well logging operations

Requirement	Training Criteria
10 CFR 39.61(a)	Logging Supervisor
B. On-the-Job Training - using sealed sources 160 hours for a mineral logging licensee, or a licensee using sealed sources with activities less than 500 millicuries **OR** 3 months or 520 hours for gas or oil well logging operations using sealed sources with activities greater than 500 millicuries	Under the supervision of a qualified logging supervisor
C. On-the-Job Training - using tracer materials **Single Well Tracer Operations** 3 months or 520 hours or completion of 50 tracer operations **Field Flood Operations** 3 months or 520 hours or completion of 3 field flood tracer operations involving multiple wells	Under the supervision of a qualified logging supervisor
D. Completion of a Written Examination	Complete a written examination submitted and approved by NRC
E. Must receive Copies of and Instruction in: (Classroom Training - Approximately 8 hours in length)	NRC Regulations • Applicable sections of 10 CFR Parts 19, 20, and 39 • The NRC License under which the logging supervisor will perform well logging • The Licensee's Operating & Emergency Procedures required by 10 CFR 39.63
F. Pass Written Examination on 10 CFR 39.61(e) Topics outlined in Item A	• Complete a written examination submitted and approved by NRC. • Passing Grade 80%

Requirement	Training Criteria
10 CFR 39.61(a)	Logging Supervisor
G. Receive Equipment Training (Approximately 4 hours in length)	Training includes: • Well Logging Equipment • Sealed sources • Handling equipment • Survey meters • Daily inspection
H. Demonstrate Understanding in Use of Well Logging Equipment by Passing Practical Field Exam	Questions on topics determined by the licensee Use the Well Logging Supervisor/Logging Assistant Inspection Checklist as a potential source of questions
I. Annual Refresher Training	Review the following: • Annual radiation safety program review • New procedures, equipment, or techniques • New regulations • Observations and deficiencies during audits of well logging supervisor and logging assistants and discussion of any significant incidents or accidents involving well logging • Employee questions
J. Records	To be maintained in accordance with 10 CFR 39.61(d)

Requirement	Training Criteria
10 CFR 39.61(b)	Logging Assistant
A. Must receive Copies of and Instruction in: (Classroom Training - Approximately 8 hours in length)	NRC Regulations • Applicable sections of 10 CFR Parts 19 and 20 • The Licensee's Operating & Emergency Procedures required by 10 CFR 39.63
B. Pass Oral or Written Exam	• Complete a written examination submitted and approved by NRC. • Passing Grade 80%

Requirement	Training Criteria
10 CFR 39.61(b)	Logging Assistant
C. Receive Equipment Training (Approximately 2-4 hours in length)	Training under the supervision of a qualified well logging supervisor appropriate for the logging assistant's intended job responsibilities: • Well logging equipment • Sealed sources • Handling equipment • Survey meters • Daily inspection
D. Annual Refresher Training	Review the following: • Any Significant item identified in the annual review of the Radiation Safety Program • New procedures or equipment • New regulations • Observations and deficiencies during audits and discussion of any significant incidents or accidents involving well logging operations • Employee questions
E. Records	To be maintained in accordance with 10 CFR 39.61(d)

Appendix M

Annual Internal Job Performance Inspection Checklist for Well Logging Supervisors and Well Logging Assistants

Annual Internal Job Performance Inspection Checklist for Well Logging Supervisors and Well Logging Assistants

Well Logging Location _____ Date _____

Logging Supervisor _____ Time _____

Logging Assistant _____

Inspector _____

Yes	No	Questions
		1. Film, TLD, or OSL badge available and properly worn?
		2. Individuals working within the restricted area wearing TLD, OSL, or film badges or dosimeters?
		3. Restricted areas properly controlled to prevent unauthorized entry?
		4. Calibrated and properly operating survey meter and evidence of its latest calibration available?
		5. Latest survey records as required by paragraphs 10 CFR 39.67(b), (c), and (e) available? • Measurements of positions occupied in transport vehicle • Measurement of vehicle exterior • Contamination check of well logging tool prior to transport • Measurements before and after subsurface tracer use
		9. Shipping papers for transportation of radioactive material available and properly filled out?
		10. Utilization log properly filled out?
		11. Defective well logging equipment being used?
		12. Copy of the applicant's operating and emergency procedures available at the site?
		13. Radioactive isotopes stored and secured properly to prevent unauthorized removal?
		14. Storage area properly posted with "Caution or Danger Radioactive Material" signs?
		15. Additional items of noncompliance noted during this audit? (If any, explain, in remarks.)

Remarks:

Appendix N

Radiation Monitoring Instrument Specifications and Model Survey Instrument Calibration Program

Radiation Monitoring Instrument Specifications and Model Survey Instrument Calibration Program

Radiation Monitoring Instrument Specifications

The specifications in Table N.1 will help applicants and licensees choose the proper radiation detection equipment for monitoring the radiological conditions at their facilities.

Table N.1   Typical Survey. *Instruments used to measure radiological conditions at licensed facilities.*[2]

Portable Instruments Used for Contamination and Ambient Radiation Surveys			
Detectors	Radiation	Energy Range	Efficiency
Exposure Rate Meters	Gamma, X-ray	μR-R	N/A
Count Rate Meters			
GM	Alpha	All energies (dependent on window thickness)	Moderate
	Beta	All energies (dependent on window thickness)	Moderate
	Gamma	All energies	< 1%
NaI Scintillator	Gamma	All energies (dependent on crystal thickness)	Moderate
Plastic Scintillator	Beta	Carbon-14 or higher (dependent on window thickness)	Moderate

Stationary Instruments Used to Measure Wipe, Bioassay, and Samples from Tracer/Field Flood Study Job sites			
Detectors	Radiation	Energy Range	Efficiency
Liquid Scintillation Counter*	Alpha	All energies	High
	Beta	All energies	High
	Gamma		Moderate

---

[2]   Table adapted from The Health Physics & Radiological Health Handbook, Revised Edition, Edited by Bernard Shleien, 1992 (except for * items).

Stationary Instruments Used to Measure Wipe, Bioassay, and Samples from Tracer/Field Flood Study Job sites *(Cont'd)*			
Gamma Spectroscopy System using a (NaI)* detector	Gamma	All energies	High
Gas Proportional	Alpha	All energies	High
	Beta	All energies	Moderate
	Gamma	All energies	< 1%

Model Instrument Calibration Program

Training

Before allowing an individual to perform survey instrument calibrations, the RSO will ensure that he or she has sufficient training and experience to perform independent survey instrument calibrations.

Classroom training may be in the form of lecture, videotape, or self-study and will cover the following subject areas:

- Principles and practices of radiation protection

- Radioactivity measurements, monitoring techniques, and using instruments

- Mathematics and calculations basic to using and measuring radioactivity

- Biological effects of radiation.

Appropriate on-the-job training consists of the following:

- Observing authorized personnel performing survey instrument calibration

- Conducting survey meter calibrations under the supervision and in the physical presence of an individual authorized to perform calibrations.

Facilities and Equipment for Calibration of Dose Rate or Exposure Rate Instruments

- To reduce doses received by individuals not calibrating instruments, calibrations will be conducted in an isolated area of the facility or at times when no one else is present

- Individuals conducting calibrations will wear assigned dosimetry

- Individuals conducting calibrations will use a calibrated and operable survey instrument to ensure that unexpected changes in exposure rates are identified and corrected

Model Procedure for Calibrating Survey Instruments

A radioactive sealed source(s) used for calibrating survey instruments will:

- Approximate a point source

- Have its apparent source activity or the exposure rate at a given distance traceable by documented measurements to a standard certified to be within ± 5% accuracy by National Institutes of Standards and Technology (NIST)

- Approximate the same energy and type of radiation as the environment in which the calibrated device will be employed or develop energy curves to compensate for differing energies

- For dose rate and exposure rate instruments, the source should be strong enough to give an exposure rate of at least about 7.7×10^{-6} coulombs/kilogram/hour (30 mR/hr) at 100 cm [e.g., 3.1 gigabecquerels (85 mCi) of cesium-137 or 7.8×10^2 megabecquerels (21 mCi) of cobalt-60]

The three kinds of scales frequently used on dose or dose rate survey meters are calibrated as follows[3]:

- Linear readout instruments with a single calibration control for all scales should be adjusted at the point recommended by the manufacturer or at a point within the normal range of use. Instruments with calibration controls for each scale should be adjusted on each scale. After adjustment, the response of the instrument should be checked at approximately 20% and 80% of full scale. The instrument's readings should be within ± 15% of the conventionally true values for the lower point and ± 10% for the upper point.

- Logarithmic readout instruments, which commonly have a single readout scale spanning several decades, normally have two or more adjustments. The instrument should be adjusted for each scale according to site specifications or the manufacturer's specifications. After adjustment, calibration should be checked at a minimum of one point on each decade.

---

[3] ANSI N323A-1997, "Radiation Protection Instrumentation Test and Calibration."

Instrument readings should have a maximum deviation from the conventionally true value of no more than 10% of the full decade value.

- Meters with a digital display device shall be calibrated the same as meters with a linear scale

- Readings above 2.58 X 10^{-4} coulomb/kilogram/hour (1 R/hr) need not be calibrated, but such scales should be checked for operation and response to radiation

- The inverse square and radioactive decay laws should be used to correct changes in exposure rate due to changes in distance or source decay.

Surface Contamination Measurement Instruments[3]

- A survey meter's efficiency must be determined by using sealed sources with similar energies and types of radiation that the survey instrument will be used to measure or by developing energy curves to compensate for differing energies.

- If each scale has a calibration potentiometer, the reading should be adjusted to read the conventionally true value at approximately 80% of full scale, and the reading at approximately 20% of full scale should be observed. If only one calibration potentiometer is available, the reading should be adjusted at mid-scale on one of the scales, and readings on the other scales should be observed. Readings should be within 20% of the conventionally true value.

Model Procedures for Calibrating, Liquid Scintillation Counters, Gamma Counters, Gas Flow Proportional Counters, and Multichannel Analyzers

A radioactive sealed source used for calibrating instruments will do the following:

- Approximate the geometry of the samples to be analyzed

- Have its apparent source activity traceable by documented measurements to a standard certified to be within ± 5% accuracy by NIST.

- Approximate the same energy and type of radiation as the samples that the calibrated device will be used to measure.

Calibration

- Calibration of survey instruments used in well logging procedures for assessing dose or exposure rates must be conducted at least every 6 months or after instrument servicing

- Calibration must produce readings within ± 20% of the actual values over the range of the instrument

- Calibration of liquid scintillation counters will include quench correction.

Calibration Records

Calibration reports, for all survey instruments, should indicate the procedure used and the data obtained. The calibration record should include:

- The owner or user of the instrument

- A description of the instrument, including the manufacturer's name, model number, serial number, and type of detector

- A description of the calibration source, including the exposure rate at a specified distance or activity on a specified date

- For each calibration point, the calculated exposure rate or count rate, the indicated exposure rate or count rate, the deduced correction factor (the calculated exposure rate or count rate divided by the indicated exposure rate or count rate), and the scale selected on the instrument

- For instruments with external detectors, the angle between the radiation flux field and the detector (i.e., parallel or perpendicular)

- For instruments with internal detectors, the angle between radiation flux field and a specified surface of the instrument

- For detectors with removable shielding, an indication whether the shielding was in place or removed during the calibration procedure

- The exposure rate or count rate from a check source, if used

- The name of the person who performed the calibration and the date it was performed.

The following information should be attached to the instrument as a calibration sticker or tag:

- For exposure rate meters, the source isotope used to calibrate the instrument (with correction factors) for each scale

- The efficiency of the instrument, for each isotope the instrument will be used to measure (if efficiency is not calculated before each use)

- For each scale or decade not calibrated, an indication that the scale or decade was checked only for function but not calibrated

- The date of calibration and the next calibration due date

- The apparent exposure rate or count rate from the check source, if used.

References: See the Notice of Availability on the inside front cover of this report to obtain a copy of:

1. Draft Regulatory Guide FC 413-4, "Guide for the Preparation of Applications for Licenses for the Use of Radioactive Materials in Calibrating Radiation Survey and Monitoring Instruments," dated June 1985.

Additional References:

2. "The Health Physics & Radiological Health Handbook, Revised Edition," edited by Bernard Shleien, dated 1992.

3. ANSI N323A-1997, "Radiation Protection Instrumentation Test and Calibration." Copies may be obtained from the American National Standards Institute, 1430 Broadway, New York, NY 10018 or ordered electronically at the following address: <www.ansi.org>.

Appendix O

Guidance for Demonstrating that Unmonitored Individuals are Not Likely to Exceed 10 Percent of the Allowable Limits

Guidance for Demonstrating that Unmonitored Individuals are Not Likely to Exceed 10 Percent of the Allowable Limits

Dosimetry is required for individual adults who are likely to receive in 1 year an occupational dose from sources external to the body in excess of 10% of the applicable regulatory limits in 10 CFR 20.1201. However, logging supervisors or logging assistants are required by 10 CFR 39.65(a) to wear either a film badge or a thermoluminescent dosimeter (TLD) when handling licensed tracer materials or sealed sources. In instances where pocket chambers are used instead of film badges or TLDs to assess radiation dosage of personnel who are not logging supervisors or logging assistants, a check of the response of the dosimeters to radiation should be made every 12 months. Acceptable pocket dosimeters should read within plus or minus 20% of the true radiation dose. To demonstrate to the NRC that dosimetry is *not* required for non-logging personnel, a licensee needs to have available an evaluation demonstrating that these nonmonitored workers are not likely to exceed 10% of the applicable annual limits — 5 mSv (500 millirems) per year.

The applicable TEDE (whole body) limit is 50 mSv (5 rems) per year, and 10% of that value is 5 mSv (500 millirems) per year.

Three common ways that individuals may exceed 10% of the applicable limits are mishandling tracer radioisotopes, logging tools, or any devices containing sealed sources. However, most routine well logging or tracer activities result in minimal doses to well logging and tracer personnel. A licensee will need to conduct an evaluation of doses occupationally exposed workers could receive in performing tasks involving the handling of radioactive materials to assess the need for dosimetry.

Example: A careful radiation measurement using a survey meter of the location producing the highest dose rate at the rear of the logging truck where radioactive material is stored in its transport compartment and where mechanics routinely work, is found to be 0.015 mSv/hr (1.5 mrem/hr). Mechanics are not expected to spend any more than a total of 3 hours per week at the location near the storage containers where the sealed sources are housed at the rear of the truck. Based on this measured dose rate, the annual dose is expected to be less than 2.34 mSv (234 mrem). Specifically, 3 hr/wk x 1.5 mrem/hr x 52 wk/yr = 234 mrem. Based on the above, if any mechanic works in the area less than 6.4 hours per week, no dosimetry is required.

Note: 6.4 hours is the total amount of hours it would take for an individual to meet the 5 mSv (500 millirems) per year limit.

Appendix P

Guidance for Demonstrating that Individual Members of the Public will not Receive Doses Exceeding the Allowable Limits

Guidance for Demonstrating that Individual Members of the Public will not Receive Doses Exceeding the Allowable Limits

Licensees must ensure that:

- The radiation dose received by individual members of the public does not exceed 1 mSv (100 mrem) in one calendar year resulting from the licensee's possession and/or use of licensed materials.

Members of the public include persons who live, work, or may be near locations where licensed material is used or stored and employees whose assigned duties do not include the use of licensed materials and who work in the vicinity where it is used or stored.

- The radiation dose in unrestricted areas does not exceed 0.02 mSv (2 mrem) in any one hour.

Typical unrestricted areas may include offices, shops, laboratories, areas outside buildings, property, and nonradioactive equipment storage areas. The licensee does not control access to these areas for purposes of controlling exposure to radiation or radioactive materials; however, the licensee may control access to these areas for other reasons, such as security.

Licensees must demonstrate compliance with both of the above regulations. For areas adjacent to facilities where licensed material is used or stored, calculations or a combination of calculations and measurements (e.g., using an environmental TLD) are often used to show compliance.

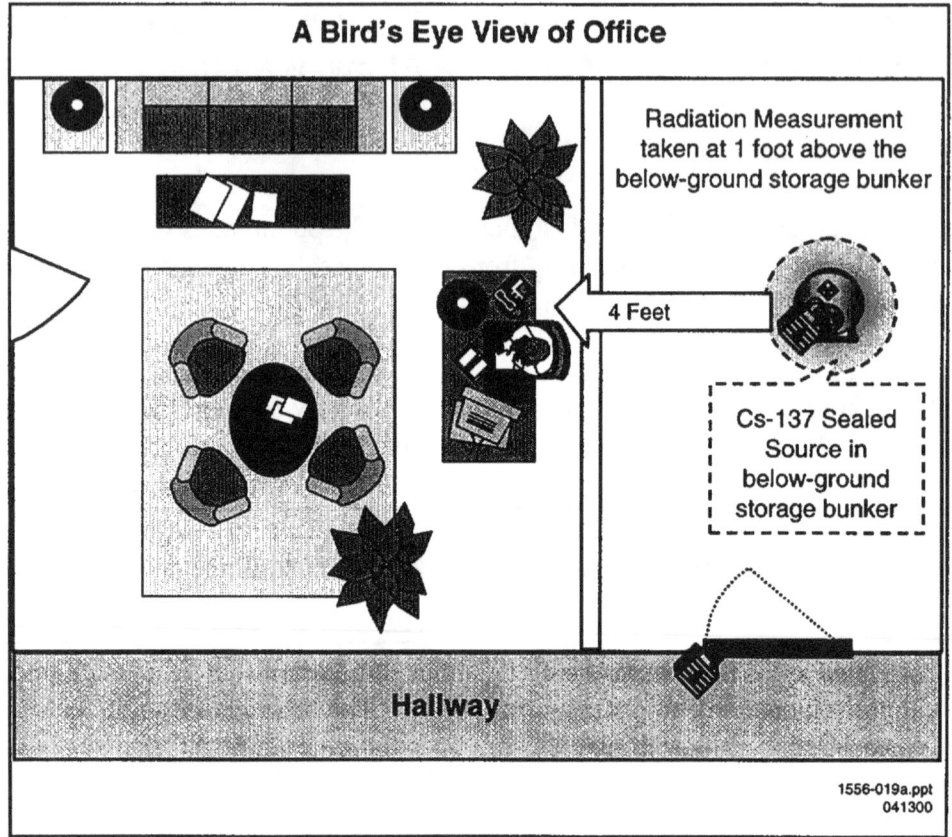

Figure P.1 Bird's Eye View of Office.

Calculation Method[4]

These measurements must be made with calibrated survey meters sufficiently sensitive to measure background levels of radiation. However, licensees must exercise caution when making these measurements, and they must use currently calibrated radiation survey instruments. A maximum dose of 1 mSv (100 mrem) received by an individual over a period of 2080 hours (i.e., a "work year" of 40 hr/wk for 52 wk/yr) is equal to less than 0.5 microsievert (0.05 mrem) per hour.

> This rate is well below the minimum sensitivity of most commonly available G-M survey instruments.

Instruments used to make measurements for calculations must be sufficiently sensitive. An instrument equipped with a scintillation-type detector (e.g., NaI(Tl)) or a micro-R meter used in making very low gamma radiation measurements should be adequate.

---

[4] For ease of use, the examples in this Appendix use conventional units. The conversions to SI units are as follows: 1 foot (ft) = 0.305 meter; 1 mrem = 0.01 mSv.

Licensees may also choose to use environmental TLDs in unrestricted areas next to the down-hole source storage area for monitoring. This direct measurement method would provide a definitive measurement of actual radiation levels in unrestricted areas without any restrictive assumptions. Records of these measurements can then be evaluated to ensure that rates in unrestricted areas do not exceed the 1 mSv/yr (100 mrem/yr) limit.

TLDs used for personnel monitoring (e.g., LiF) may not have sufficient sensitivity for this purpose. Generally, the minimum reportable dose received is 0.1 mSv (10 mrem). Suppose a TLD monitors dose received and is changed once a month. If the measurements are at the minimum reportable level, the annual dose received could have been about 1.2 mSv (120 mrem), a value in excess of the 1 mSv/yr (100 mrem/yr) limit. If licensees use TLDs to evaluate compliance with the public dose limits, they should consult with their TLD supplier and choose more sensitive TLDs, such as those containing CaF_2 that are used for environmental monitoring.

The combined measurement-calculational method may be used to estimate the maximum dose to a member of the public. The combined measurement-calculational method takes a tiered approach, going through a two-part process, starting with a worst case situation and moving toward more realistic situations. It makes the following simplifications: (1) each cesium-137 logging source is a point source; (2) typical radiation levels are encountered when the source is in the unshielded position; and (3) no credit is taken for any shielding found between the source storage area and the unrestricted areas. The method is only valid for the source activity at the time of measurement and must be repeated if the source strength or shielding is changed.

Part 1 of the combined measurement-calculational method is simple but conservative. It assumes that an affected member of the public is present 24 hours a day and uses only the inverse square law to determine if the distance between the down-hole storage area and the affected member of the public is sufficient to show compliance with the public dose limits. Part 2 considers not only distance, but also the time that the affected member of the public is actually in the area under consideration. Using this approach, licensees make only those calculations that are needed to demonstrate compliance. The results of these calculations typically result in higher radiation levels than would exist at typical facilities, but they provide a method for estimating conservative doses that could be received.

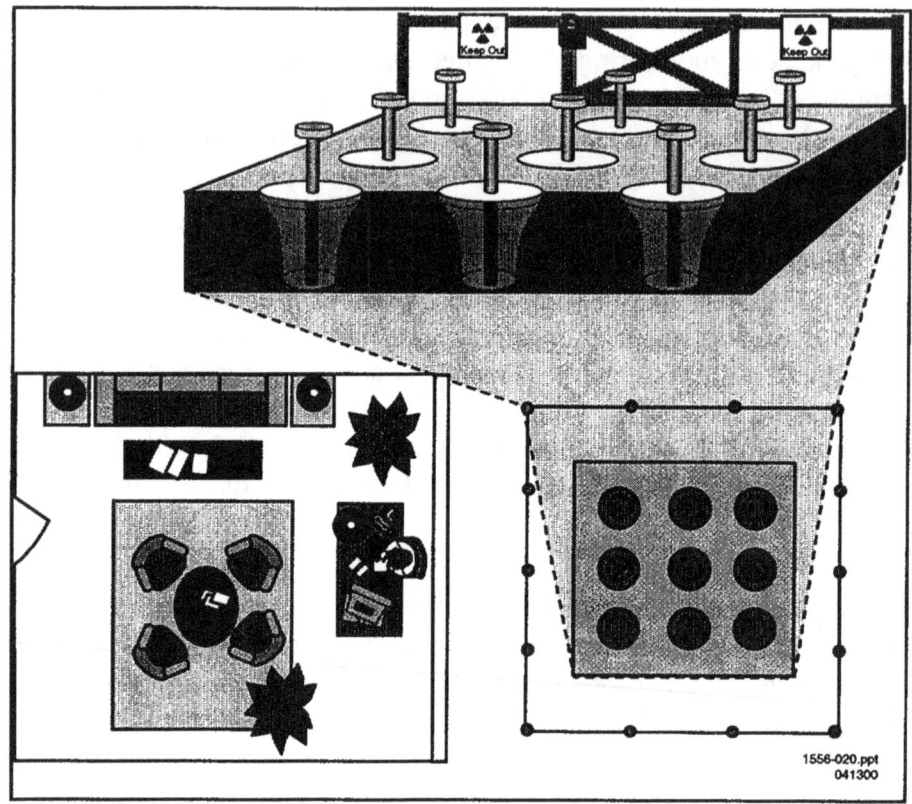

Figure P.2 Downhole Storage Array.

Example

To better understand the combined measurement-calculational method, we will examine EZ Well Logging, Inc., a well logging licensee. Yesterday, the company's president noted that the top shield of the down-hole storage area is close to an area used by workers whose assigned duties do not include the use of licensed materials, and he asked Elmo, the Radiation Safety Officer (RSO), to determine if the company is complying with NRC's regulations.

The area in question is near the floor under the workers' desks, which constitutes the primary shield of the down-hole storage area. Elmo measures the distance from the shield to the center of the area in question and, using a calibrated survey instrument, measures the highest dose rate at one foot from the shield to be 2 mrem per hour.

Table P.1 summarizes the information Elmo has on the down-hole storage area.

Table P.1   Information Known about Dose at the Shield of the Cs-137 Source

Description of Known Information	Cesium-137 Logging Source
Dose rate encountered at 1 foot from the top of the shield, in mrem/hr.	2 mrem/hr.
Distance from the face of the shield to the nearest occupied work area, in ft.	4 ft

Example: Part 1

Elmo's first thought is that the distance between the down-hole storage area shield and the area in question may be sufficient to show compliance with the regulation in 10 CFR 20.1301. So, taking a worst case approach, he assumes: 1) the cesium-137 is constantly located in down-hole storage area (i.e., 24 hr/d), and 2) the workers are constantly in the unrestricted work area (i.e., 24 hr/d). Elmo proceeds to calculate the dose the workers might receive hourly and yearly from the source, as shown in Table P-2 below.

Table P.2   Calculational Method, Part 1: Hourly and Annual Doses Received from a Logging Source Stored in Above Ground Transportation Container.

Step No.	Description	Input Data	Results
1	Multiply the measured dose rate measured at 1.0 ft from the face of the shield floor in mrem/hr by the square of the distance (ft) at which the measurement was made (e.g., 1 foot from the face of the shield)	$2 \times (1)^2$	2
2	Square of the distance (ft) from the face of the shield to the nearest unrestricted area, in ft^2	$(4)^2$	16
3	Divide the result of Step 1 by the result of Step 2 to calculate the dose received by an individual in the area near the shield. **HOURLY DOSE RECEIVED FROM SOURCE**, in mrem in an hour	2/16	**0.125**
4	Multiply the result of Step 3 by 40 hr/work week x 52 weeks/year = **MAXIMUM ANNUAL DOSE RECEIVED FROM Cs-137 Source**, in mrem in a year	0.125 X 40 X 52	**260**

Note: The result in Step 3 demonstrates compliance with the 2 mrem in any one hour limit. Re-evaluate if assumptions change. If the result in Step 4 exceeds 100 mrem/yr, proceed to Part 2 of the calculational method.

At this point, Elmo is pleased to see that the total dose that an individual could receive in any one hour is only 0.125 mrem in an hour, less than the 2 mrem in any one hour limit but notes that an individual could receive a dose of 260 mrem in a year, higher than the 100 mrem limit.

Example: Part 2

Elmo reviews the assumptions and recognizes that the workers are not in area near the shield all of the time. A realistic estimate of the number of hours the workers spend in the area is made, keeping the other assumptions constant (i.e., the source is constantly in the down-hole storage area (i.e., 24 hr/d). The annual dose received is then recalculated.

Table P.3   Calculational Method, Part 2: Annual Dose Received from a Logging Source Stored in Above Ground Transportation Container.

Step No.	Description	Results
7	A. Average number of hours per day an individual spends in area of concern (e.g., a non-radiation worker spends 1.5 hr/day in the area near the shield; the remainder of the day the workers are away from the area assigned to jobs unrelated to radiation. (painting, grounds keeping, desk jobs, etc.) B. Average number of days per week in area C. Average number of weeks per year in area (e.g., full time workers)	1.5552
8	Multiply the results of Step 7.A. by the results of Step 7.B. by the results of Step 7.C. = **AVERAGE NUMBER OF HOURS IN AREA OF CONCERN PER YEAR**	1.5 x 5 x 52 = 390
9	Multiply the results in Step 3 by the results of Step 8 = **ANNUAL DOSE RECEIVED FROM CESIUM-137 LOGGING SOURCE CONSIDERING REALISTIC ESTIMATE OF TIME SPENT IN AREA OF CONCERN**, in mrem in a year	0.125 x 390 = 49

Elmo is pleased to note that the calculated annual dose received is significantly lower, and does not exceed the 100 mrem in a year limit.

Elmo is glad to see that the results in Step 9 show compliance with the 100 mrem in a year limit. Had the result in Step 9 been higher than 100 mrem in a year, then Elmo could have done one or more of the following:

- Consider whether the assumptions used to determine occupancy are accurate, revise the assumptions as needed, and recalculate using any new assumptions

- Calculate the effect of any shielding located between the storage area and the floor of the public area — such calculation is beyond the scope of this Appendix

- Take corrective action (e.g., change work patterns to reduce the time spent in the area near the shield) and perform new calculations to demonstrate compliance

- Designate the area inside the use area as a restricted area and the workers as occupationally exposed individuals. This would require controlling access to the area for purposes of radiation protection and training the workers as required by 10 CFR 19.12.

National Council on Radiation Protection and Measurements (NCRP) Report No. 49, "Structural Shielding Design and Evaluation for Medical Use of X Rays and Gamma Rays of Energies Up to 10 MeV," contains helpful information. It is available from NCRP, 7910 Woodmont Avenue, Suite 800, Bethesda, Maryland 20814. NCRP's telephone numbers are: (301) 657-2652 or 1-800-229-2652.

Note that in the example, Elmo evaluated the unrestricted area outside only one wall of the down-hole storage area. Licensees also need to make similar evaluations for other unrestricted areas and to keep in mind the ALARA principle, taking reasonable steps to keep radiation dose received below regulatory requirements. In addition, licensees need to be alert to changes in situations (e.g., adding sources to the storage area, changing the work habits of the workers, or otherwise changing the estimate of the portion of time spent in the area in question) and to perform additional evaluations, as needed.

RECORD KEEPING: 10 CFR 20.2107 requires licensees to maintain records demonstrating compliance with the dose limits for individual members of the public.

Notifications

Event	Telephone Notification	Written Report	Regulatory Requirement
Theft or loss of material	immediate	30 days	10 CFR 20.2201(a)(1)(i)
Whole body dose greater than 0.25 Sv (25 rems)	immediate	30 days	10 CFR 20.2202(a)(1)(i)
Extremity dose greater than 2.5 Sv (250 rems)	immediate	30 days	10 CFR 20.2202(a)(1)(iii)
Whole body dose greater than 0.05 Sv (5 rems) in 24 hours	24 hours	30 days	10 CFR 20.2202(b)(1)(i)
Extremity dose greater than 0.5 Sv (50 rems) in 24 hours	24 hours	30 days	10 CFR 20.2202(b)(1)(iii)
Whole body dose greater than 0.05 Sv (5 rems)	none	30 days	10 CFR 20.2203(a)(2)(i)
Dose to individual member of public greater than 1 mSv (100 mrems)	none	30 days	10 CFR 20.2203(a)(2)(iv)
Defect in equipment that could create a substantial safety hazard subject to the requirements of 10 CFR Parts 30, 40, and 70	2 days	30 days	10 CFR 21.21(d)(3)(i)
Event that prevents immediate protective actions necessary to avoid exposure to radioactive materials that could exceed regulatory limits	immediate	30 days	10 CFR 30.50(a)
Equipment is disabled or fails to function as designed when required to prevent radiation exposure in excess of regulatory limits	24 hours	30 days	10 CFR 30.50(b)(2)
Unplanned fire or explosion that affects the integrity of any licensed material or device, container, or equipment with licensed material	24 hours	30 days	10 CFR 30.50(b)(4)
Rupture of sealed source	immediate	30 days	10 CFR 39.77(a)

Event	Telephone Notification	Written Report	Regulatory Requirement
Sealed source becomes lodged in well bore and becomes classified as irretrievable, or licensee is requesting an extension to complete abandonment procedures	24 hours	30 days	10 CFR 39.77(c) 10 CFR 39.77(d)
Leak test of sealed source resulting in leakage greater than 185 Bq (0.005 microcuries)	none	5 days	10 CFR 39.35(d)
Failure of any component to perform its intended function	none	30 days	10 CFR 21.21

Note: Telephone notifications shall be made to the NRC Operations Center at **301-951-0550** except as noted.

Appendix R

Model Leak Test Program

Model Leak Test Program

Training

Before allowing an individual to perform leak test analysis independently, the RSO will ensure that this individual has sufficient classroom and on-the-job training to show competency in performing leak test analysis.

Classroom training in the performance of leak test analysis may be provided in the form of lecture, videotape, or self-study. This should cover the following subject areas:

- Principles and practices of radiation protection

- Radioactivity measurements, monitoring techniques, and using instruments

- Mathematics and calculations basic to using and measuring radioactivity

- Biological effects of radiation.

Appropriate on-the-job training consists of:

- Observing authorized personnel collecting and analyzing leak test samples

- Collecting and analyzing leak test samples under the supervision and in the physical presence of an individual authorized to perform leak tests and leak test analysis

Facilities and Equipment

- To ensure the required sensitivity of measurements, leak tests will be analyzed in a low-background area.

- Before leak test swipes are analyzed, individuals conducting leak tests will use a calibrated and operable survey instrument to check leak test samples for gross contamination. If the sensitivity of the counting system is unknown, the minimum detectable activity (MDA) needs to be determined. The MDA may be determined using the following formula:

$$MDA = \frac{3 + 4.65(BR)^{*\frac{1}{2}}}{Et}$$

where MDA = activity level in disintegrations per minute (dpm)
　　　　BR = background rate in counts per minute (cpm)
　　　　t = counting time in minutes
　　　　E = detector efficiency in counts per disintegration (cpd)

For example:

where BR = 200 cpm
       E   = 0.1 cpd (10% efficient)
       t   = 2 minutes

       MDA = $\dfrac{3 + 4.65(200 \text{ cpm})^{*\frac{1}{2}}}{(0.1 \text{ cpd})(2 \text{ minutes})}$

A NaI(Tl) well counter system with a single or multi-channel analyzer will be used to count samples from sealed sources containing gamma-emitters (e.g., cesium-137, cobalt-60).

A liquid scintillation, gas-flow proportional, or solid state counting system will be used to count samples containing alpha-emitters (e.g., americium-241).

Frequency for Conducting Leak Tests of Sealed Sources

Leak tests on well logging sealed sources will be conducted at intervals not to exceed 6 months, or, for ECSs requiring leak tests, at intervals not to exceed 3 years.

Procedure for Performing Leak Testing and Analysis

- For each source to be tested, list identifying information such as the manufacturer's name, model number, serial number, radionuclide, and activity of the sealed source(s).

- Prepare a separate wipe sample (e.g., cotton swab or filter paper) for each source.

- Number each wipe to correlate with identifying information for each source.

- If available, use a survey meter to monitor exposure.

- Wipe the most accessible area (but not directly from the surface of the source) where contamination would accumulate if the sealed source were leaking, e.g., the leak test can be taken of the part that connects to the source or the inside of the transport container that has recently transported the source.

- Select an instrument that is sensitive enough to detect 185 Bq (0.005 mCi) of the radionuclide of the sealed source.

- Using the selected instrument, count and record background count rate.

- Check the instrument's counting efficiency using a standard source of the same radionuclide as the source being tested or one with similar energy characteristics. Accuracy of standards should be within ± 5% of the stated value and traceable to primary radiation standard, such as those maintained by the National Institutes of Standards and Technology (NIST).

- Calculate efficiency.

For example: $\frac{[(cpm\ from\ std)\ -\ (cpm\ from\ bkg)]}{activity\ of\ std\ in\ Bq}$ = efficiency in cpm/Bq

where: cpm = counts per minute
       std = standard
       bkg = background
       Bq = Becquerel

- Count each wipe sample; determine net count rate.

- For each sample, calculate and record estimated activity in Bq (or μCi).

For example: $\frac{[(cpm\ from\ wipe\ sample)\ -\ (cpm\ from\ bkg)]}{efficiency\ in\ cpm/Bq}$ = Bq on wipe sample

Sign and date the list of sources, data and calculations. Retain records for 3 years (10 CFR 20.2103(a)).

If the wipe test activity is 185 Bq (0.005 microcurie) or greater, notify the RSO, so that the source can be withdrawn from use and disposed of properly. Also notify NRC.

Appendix S

Transportation - Major DOT Regulations; Sample Shipping Documents, Placards and Labels

Transportation - Major DOT Regulations; Sample Shipping Documents, Placards and Labels

The major areas in the DOT regulations that are most relevant for transportation of licensed material shipped as Type A quantities are as follows:

- Hazardous Materials Table, 49 CFR 172.101, App. A, list of hazardous substances and reportable quantities (RQ), Table 2: Radionuclides

- Shipping Papers 49 CFR 172.200-204: General entries, description, additional description requirements, shipper's certification

- Package Markings 49 CFR 172.300, 49 CFR 172.301, 49 CFR 172.303 49 CFR 172.304, 49 CFR 172.310, 49 CFR 172.324: General marking requirements for non-bulk packaging, prohibited marking, marking requirements, radioactive material, hazardous substances in non-bulk packaging

- Package Labeling 49 CFR 172.400, 49 CFR 172.401, 49 CFR 172.403, 49 CFR 172.406, 49 CFR 172.407, 49 CFR 172.436, 49 CFR 172.438, 49 CFR 172.440: General labeling requirements, prohibited labeling, radioactive materials, placement of labels, specifications for radioactive labels

- Placarding of Vehicles 49 CFR 172.500, 49 CFR 172.502, 49 CFR 172.504, 49 CFR 172.506, 49 CFR 172.516, 49 CFR 172.519, 49 CFR 172.556: Applicability, prohibited and permissible placarding, general placarding requirements, providing and affixing placards: highway, visibility and display of placards, specifications for RADIOACTIVE placards

- Emergency Response Information, Subpart G, 49 CFR 172.600, 49 CFR 172.602, 49 CFR 172.604: Applicability and general requirements, emergency response information, emergency response telephone number

- Training, Subpart H, 49 CFR 172.702, 49 CFR 172.704: Applicability and responsibility for training and testing, training requirements

- Shippers - General Requirements for Shipments and Packaging, Subpart I, 49 CFR 173.403, 49 CFR 173.410, 49 CFR 173.412, 49 CFR 173.415, 49 CFR 173.431, 49 CFR 173.433, 49 CFR 173.435, 49 CFR 173.441, 49 CFR 173.443, 49 CFR 173.448, 49 CFR 173.475, 49 CFR 173.476: Definitions, general design requirements, additional design requirements for Type A packages, authorized Type A packages, activity limits for Type A packages, requirements for determining A_1 and A_2 values, table of A_1 and A_2 values for radionuclides, radiation level limitations, contamination control, general transportation requirements, quality control requirements prior to each shipment, approval of special form radioactive materials

- Carriage by Public Highway - General Information and Regulations, Subpart A, 49 CFR 177.816, 49 CFR 177.817, 49 CFR 177.834(a), 49 CFR 177.842: Driver training, shipping paper, general requirements (secured against movement), Class 7 (radioactive) material.

The following are the major areas in DOT regulations most relevant for transporting licensed material that is shipped as Type B quantities in addition to the applicable requirements stated above:

A. Package Markings
   49 CFR 172.310 - Radioactive material [Type B]

B. Shippers - General Requirements for Shipments and Packaging - 49 CFR 173

   1. 49 CFR 173.25 - Requirements for use and labeling of overpacks

   2. 49 CFR 173.403 - Definitions

   3. 49 CFR 173.411 - General design requirements

   4. 49 CFR 173.413 - Additional design requirements for Type B packages

   5. 49 CFR 173.416 - Authorized Type B packages [includes packaging certification requirements]

   6. 49 CFR 173.471 - Additional requirements for Type B packages approved by NRC

Part 2

Sample Shipping Documents, Placards and Labels

Hazard Communications for Class 7 (Radioactive) Materials

DOT Shipping Papers (49 CFR 172.200-205)

NOTE: IAEA, ICAO, and IMO may require additional hazard communication information for international shipments
This table must not be used as a substitute for the DOT and NRC regulations on the transportation of radioactive materials

Entries Always Required Unless Excepted	Additional Entries Sometimes Required	Optional Entries
• The basic description, in sequence: **Proper Shipping Name, Hazard Class (7), U.N. Identification Number** • 24 hour **emergency response telephone number** • Name of **shipper** • Proper page numbering (Page 1 of 4) • Except for empty and bulk packages, the **total quantity** (mass, or volume for liquid), in appropriate units (lbs, mL....) • If not special form, **chemical and physical form** • The **name of each radionuclide** (95% rule) and total package activity. The activity must be in SI units (e.g., Bq, TBq), or both SI units and customary units (e.g., Ci, mCi). However, for <u>domestic</u> <u>shipments</u>, the activity *may be* expressed in terms of customary units only, until 4/1/97. • For each labeled package: ‐ The **category of label** used; ‐ The **transport index** of each package with a Yellow-II or Yellow-III label • Shipper's **certification** (not required of private carriers)	<u>Materials-Based</u> <u>Requirements</u>: • If hazardous substance, "RQ" as part of the basic description • The LSA or SCO group (e.g., LSA-II) • "Highway Route Controlled Quantity" as part of the basic description, if HRCQ • Fissile material information (e.g., "Fissile Exempt," controlled shipment statement [see §172.203(d)(7)]) • If the material is considered hazardous waste and the word waste does not appear in the shipping name, then "waste" must precede the shipping name (e.g., Waste Radioactive Material, nos, UN2982) • "Radioactive Material" if not in proper shipping name <u>Package-Based</u> <u>Requirements</u>: • Package identification for DOT Type B or NRC certified packages • IAEA CoC ID number for export shipments or shipments using foreign-made packaging (see §173.473) <u>Administrative-Based</u> <u>Requirements</u>: • "Exclusive Use-Shipment" • Instructions for maintenance of exclusive use-shipment controls for LSA/SCO strong-tight or NRC certified LSA (§ 173.427) • If a DOT exemption is being used, "DOT-E" followed by the exemption number	• The type of packaging (e.g., Type A, Type B, IP-1, ...) • The Technical/chemical name may be included (if listed in §172.203(k), in parentheses between the proper shipping name and hazard class; otherwise inserted in parenthesis after the basic description) • Other information is permitted (e.g., functional description of the product), provided it does not confuse or detract from the proper shipping name or other required information • For fissile radionuclides, except Pu-238, Pu-239, and Pu-241, the weight in grams or kilograms may be used *in place of* activity units. For Pu-238, Pu-239, and Pu-241, the weight in grams or kilograms may optionally be entered *in addition to* activity units [see § 172.203(d)(4)] • Emergency response hazards and guidance information (§§ 172.600-604) may be entered on the shipping papers, or may be carried with the shipping papers [§ 172.602(b)]

Some Special Considerations/Exceptions for Shipping Paper Requirements

• Shipments of Radioactive Material, excepted packages, under UN2910 (e.g., Limited Quantity, Empty packages, and Radioactive Instrument and Article), are excepted from shipping papers. For limited quantities (§173.421), this is only true if the limited quantity is not a hazardous substance (RQ) or hazardous waste (40 CFR 262)

• Shipping papers must be in the pocket on the left door, or readily visible to person entering driver's compartment and within arm's reach of the driver

• For shipments of multiple cargo types, any HAZMAT entries must appear as the first entries on the shipping papers, be designated by an "X" (or "RQ") in the hazardous material column, <u>or</u> be highlighted in a contrasting color

Hazard Communications for Class 7 (Radioactive) Materials

Marking Packages (49 CFR 172.300-338)

NOTE: IAEA, ICAO, and IMO may require additional hazard communication information for international shipments
This table must not be used as a substitute for the DOT and NRC regulations on the transportation of radioactive materials

Markings Always Required Unless Excepted	Additional Markings Sometimes Required	Optional Markings
Non-Bulk Packages • Proper shipping name • U.N. identification number • Name and address of consignor or consignee, *unless*: - highway only and no motor carrier transfers, or - part of carload or truckload lot or freight container load, and entire contents of railcar, truck, or freight container are shipped from one consignor to one consignee [see §172.301(d)] - - - - - - - - - - - - - - - - Bulk Packages (i.e., net capacity greater than 119 gallons as a receptacle for liquid, or 119 gallons and 882 pounds as a receptacle for solid, or water capacity greater than 1000 lbs, with no consideration of intermediate forms of containment) • U.N. identification number, on orange, rectangular panel (see §172.332) - some exceptions exist	Materials-Based Requirements: • If in excess of 110 lbs (50 kg), Gross Weight • If non-bulk liquid package, underlined double arrows indicating upright orientation (two opposite sides) [ISO Std 780-1985 marking] • If a Hazardous substance in non-bulk package, the letters "RQ" in association with the proper shipping name Package-Based Requirements: • The package type if Type A or Type B (½" or greater letters) • The specification-required markings [e.g., for Spec. 7A packages: "DOT 7A Type A" and "Radioactive Material" (see §178.350-353)] • For approved packages, the certificate ID number (e.g., USA/9166/B(U), USA/9150/B(U)-85, ...) • If Type B, the trefoil (radiation) symbol per Part 172 App. B [*size*: outer radius ≥ 20 mm (0.8 in)] • For NRC certified packages, the model number, gross weight, and package ID number (10 CFR 71.85) Administrative-Based Requirements: • If a DOT exemption is being used, "DOT-E" followed by the exemption number • If an export shipment, "USA" in conjunction with the specification markings or certificate markings	• "IP-1," "IP-2," or "IP-3" on industrial packaging is recommended • Both the name and address of consignor and consignee are recommended • Other markings (e.g., advertising) are permitted, but must be sufficiently away from required markings and labeling

Some Special Considerations/Exceptions for Marking Requirements

• Marking is required to be: (1) durable, (2) printed on a package, label, tag, or sign, (3) unobscured by labels or attachments, (4) isolated from other marks, and (5) be representative of the hazmat contents of the package

• Limited Quantity (§173.421) packages and Articles Containing Natural Uranium and Thorium (§173.426) must bear the marking "radioactive" on the outside of the inner package or the outer package itself, and are excepted from other marking. The excepted packages shipped under UN 2910 must also have the accompanying statement that is required by §173.422.

• Empty (§173.428) and Radioactive Instrument and Article (§173.424) packages are excepted from marking

• Shipment of LSA or SCO required by §173.427 to be consigned as exclusive use are excepted from marking except that the exterior of each nonbulk package must be marked **"Radioactive-LSA"** or **"Radioactive-SCO,"** as appropriate. Examples of this category are domestic, strong-tight containers with less than an A_2 quantity, and domestic NRC certified LSA/SCO packages using 10 CFR 71.52.

• For bulk packages, marking may be required on more than one side of the package (see 49 CFR 172.302(a))

Hazard Communications for Class 7 (Radioactive) Materials

Labeling Packages (49 CFR 172.400-450)

NOTE: IAEA, ICAO, and IMO may require additional hazard communication information for international shipments
This table must not be used as a substitute for the DOT and NRC regulations on the transportation of radioactive materials

Placement of Radioactive Labels

- Labeling is required to be: (1) placed near the required marking of the proper shipping name, (2) printed or affixed to the package surface (not the bottom), (3) in contrast with its background, (4) unobscured by markings or attachments, (5) within color, design, and size tolerance, and (6) representative of the HAZMAT contents of the package

- For labeling of radioactive materials packages, two labels are required on opposite sides excluding the bottom

Determination of Required Label

Size:				
Sides: ≥ 100 mm (3.9 in.) Border: 5-6.3 mm (0.2-0.25 in.)	RADIOACTIVE I 7 — 49 CFR 172.436	RADIOACTIVE II 7 — 49 CFR 172.438	RADIOACTIVE III 7 — 49 CFR 172.440	EMPTY (6 inches × 6 in.) — 49 CFR 172.450
Label	**WHITE-I**	**YELLOW-II**	**YELLOW-III**	**EMPTY LABEL**
Required when:	Surface radiation level < 0.005 Mev/hr (0.5 mem./hr)	0.005 Mev/hr (0.5 mem./hr) < surface radiation level ≤ 0.5 Mev/hr (50 mem./hr)	0.5 Mev/hr (50 mem./hr) < surface radiation level < 2 Mev/hr (200 mem./h) [Note: 10 Mev/hr (1000 mem./hr) for exclusive-use closed vehicle (§173.441(b)]	The EMPTY label is required for shipments of empty Class 7 (radioactive) packages made pursuant to §173.428. It must cover any previous labels, or they must be removed or obliterated.
Or:	TI = 0 [1 meter dose rate < 0.0005 Mev/hr (0.05 mem./hr)]	TI ≤ 1 [1 meter dose rate < 0.01 Mev/hr (1 mem./hr)]	TI ≤ 10 [1 meter dose rate < 0.1 Mev/hr (10 mem./hr)] [Note: There is no *package* TI limit for exclusive-use]	

Notes:
- Any package containing a Highway Route Controlled Quantity (HRCQ) must bear YELLOW-III label
- Although radiation level transport indices (TIs) are shown above, for **fissile material**, the TI is typically determined on the basis of criticality control

Content on Radioactive Labels

- RADIOACTIVE Label must contain (entered using a durable, weather-resistant means):
  (1) The radionuclides in the package (with consideration of available space). Symbols (e.g., Co-60) are acceptable
  (2) The activity in SI units (e.g., Bq, TBq), or both SI units with customary units (e.g., Ci, mCi) in parenthesis. However, for domestic shipments, the activity *may* be expressed in terms of customary units only, until 4/1/97.
  (3) The Transport Index (TI) in the supplied box. The TI is entered *only* on YELLOW-II and YELLOW-III labels

Some Special Considerations/Exceptions for Labeling Requirements

- For materials meeting the definition of another hazard class, labels for each secondary hazard class need to be affixed to the package. The subsidiary label *may* not be required on opposite sides and must not display the hazard class number

- Radioactive Material, excepted packages, under UN2910 (e.g., Limited Quantity, Empty packages, and Radioactive Instrument and Article), are excepted from labeling. However, if the excepted quantity meets the definition for another hazard class, it is re-classed for that hazard. Hazard communication requirements for the other class are required

- Labeling exceptions exist for shipment of LSA or SCO required by § 173.427 to be consigned as exclusive use

- The "Cargo Aircraft Only" label is typically required for radioactive materials packages shipped by air [§ 172.402(c)]

Hazard Communications for Class 7 (Radioactive) Materials

Placarding Vehicles (49 CFR 172.500-560)

NOTE: IAEA, ICAO, and IMO may require additional hazard communication information for international shipments
This table must not be used as a substitute for the DOT and NRC regulations on the transportation of radioactive materials

Visibility and Display of Radioactive Placard

- Placards are required to be displayed:
  - on four sides of the vehicle
  - visible from the direction they face (for the front side of trucks, tractor-front, trailer, or both are authorized)
  - clear of appurtenances and devices (e.g., ladders, pipes, tarpaulins)
  - at least 3 inches from any markings (such as advertisements) which may reduce placard's effectiveness
  - upright and on-point such that the words read horizontally
  - in contrast with the background, or have a lined-border which contrasts with the background
  - such that dirt or water from the transport vehicle's wheels will not strike them
  - securely attached or affixed to the vehicle, or in a holder.

- Placard must be maintained by carrier to keep color, legibility, and visibility.

Conditions Requiring Placarding

- Placards are required for any vehicle containing package with a RADIOACTIVE Yellow-III label

- Placards are required for shipment of LSA or SCO required by §173.427 to be consigned as exclusive use. Examples of this category are domestic, strong-tight containers with less than an A_2 quantity, and domestic NRC certified LSA/SCO packages using 10 CFR 71.52. Also, for bulk packages of these materials, the orange panel *marking* with the UN identification number is not required.

- Placards are required any vehicle containing package with a Highway Route Controlled Quantity (HRCQ). In this case, the placard must be placed in a square background as shown below (see §173.507(a))

Radioactive Placard

Size Specs:

Sides: $\geq$ 273 mm (10.8 in.)

Solid line inner border: About 12.7 mm (0.5 in.) from edges

Lettering: $\geq$ 41 mm (1.6 in.)

Square for HRCQ: 387mm (15.25 in.) outside length by 25.4 mm (1 in.) thick

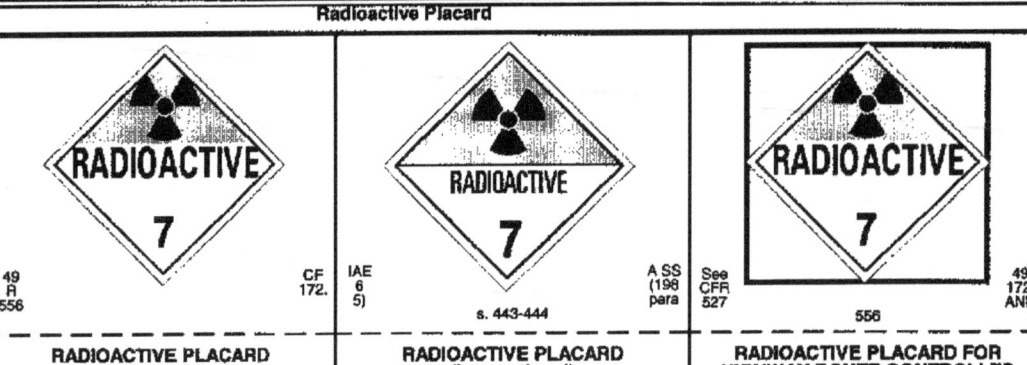

RADIOACTIVE PLACARD (Domestic) *Base of yellow solid area:* 29 ± 5 mm (1.1 + 0.2 in.) above horizontal centerline	**RADIOACTIVE PLACARD (International)**	**RADIOACTIVE PLACARD FOR HIGHWAY ROUTE CONTROLLED QUANTITY** (either domestic or international placard could be in middle)

Some Special Considerations/Exceptions for Placarding Requirements

- Domestically, substitution of the UN ID number for the word "RADIOACTIVE" on the placard is prohibited for Class 7 materials. However, some import shipments may have this substitution in accordance with international regulations.

- Bulk packages require the orange, rectangular panel marking containing the UN ID number, which must be placed adjacent to the placard (see §172.332) [NOTE: except for LSA/SCO exclusive use under §173.427, as above]

- If placarding for more than one hazard class, subsidiary placards must not display the hazard class number. Uranium Hexaflouride (UF_6) shipments $\geq$ 454 kg (1001 lbs) require both RADIOACTIVE and CORROSIVE (Class 8) placarding

- For shipments of radiography cameras in convenience overpacks, if the overpack does not require a RADIOACTIVE - YELLOW III label, vehicle placarding is not required (regardless of the label which must be placed on the camera)

Minimum Required Packaging For Class 7 (Radioactive) Materials

This table must not be used as a substitute for the DOT and NRC regulations on the transportation of radioactive materials

Quantity:	< 70 Bq/g (< 0.002 µCi/g)	Limited Quantity (§173.421)	A_1/A_2 value (§173.435)	1 rem/hr at 3 m, unshielded (§173.427)
Non-LSA/SCO:	Excepted	Type A		Type B [3]
Domestic or International LSA/SCO: LSA-I solid, (liquid)[1] SCO-I	Excepted	IP-I		Type B [3]
LSA-I Liquid LSA-II Solid, (liquid or gas)[1] (LSA-III)[1] SCO-II	Excepted	IP-II		Type B [3]
LSA-II Liquid or Gas LSA-III		IP-III		Type B [3]
Domestic (only) LSA/SCO: LSA-I, II, III; SCO-I, II	Excepted	Strong-tight [2]	DOT Spec. 7A Type A	Type B [3] NRC Type A LSA [3,4]

1. For entries in parentheses, exclusive use is required for shipment in an IP (e.g., shipment of LSA-I liquid in an IP-I packaging would require exclusive use consignment)
2. Exclusive use required for strong-tight container shipments made pursuant to §173.427(b)(2)
3. Subject to conditions in Certificate, if NRC package
4. Exclusive use required, see §173.427(b)(4). Use of these packages expires on 4/1/99 (10 CFR 71.52)

Package and Vehicle Radiation Level Limits (49 CFR 173.441) [A]

This table must not be used as a substitute for the DOT and NRC regulations on the transportation of radioactive materials

Transport Vehicle Use:	Non-Exclusive	Exclusive		
Transport Vehicle Type:	Open or Closed	Open (flat-bed)	Open w/Enclosure [B]	Closed
Package (or freight container) Limits:				
External Surface	2 Mev/hr (200 mem./hr)	2 Mev/hr (200 mem./hr)	10 Mev/hr (1000 mem./hr)	10 Mev/hr (1000 mem./hr)
Transport Index (TI) [C]	10	no limit		
Roadway or Railway Vehicle (or freight container) Limits:				
Any point on the outer surface	N/A	N/A	N/A	2 Mev/hr (200 mem./hr)
Vertical planes projected from outer edges	N/A	2 Mev/hr (200 mem./hr)	2 Mev/hr (200 mem./hr)	N/A
Top of . . .		load: 2 mSv/hr (200 mem./hr)	enclosure: 2 Mev/hr (200 mem./hr)	vehicle: 2 Mev/hr (200 mem./hr)
2 meters from. . .		vertical planes: 0.1 Mev/hr (10 mem./hr)	vertical planes: 0.1 Mev/hr (10 mem./hr)	outer lateral surfaces: 0.1 Mev/hr (10 mem./hr)
Underside		2 Mev/hr (200 mem./hr)		
Occupied position	N/A [D]	0.02 Mev/hr (2 mem./hr) [E]		
Sum of package TI's	50	no limit [F]		

A. The limits in this table do not apply to excepted packages - see 49 CFR 173.421-426
B. Securely attached (to vehicle), access-limiting enclosure; package personnel barriers are considered as enclosures
C. For nonfissile radioactive materials packages, the dimensionless number equivalent to maximum radiation level at 1 m (3.3 feet) from the exterior package surface, in millirem/hour
D. No dose limit is specified, but separation distances apply to Radioactive Yellow-II or Radioactive Yellow-III labeled packages
E. Does not apply to private carrier wearing dosimetry if under radiation protection program satisfying 10 CFR 20 or 49 CFR 172 Subpart I
F. Some fissile shipments may have combined conveyance TI limit of 100 - see 10 CFR 71.59 and 49 CFR 173.457

Package and Vehicle Contamination Limits (49 CFR 173.443)

This table must not be used as a substitute for the DOT and NRC regulations on the transportation of radioactive materials

NOTE: All values for contamination in DOT rules are to be averaged over each 300 cm²
Sufficient measurements must be taken in the appropriate locations to yield representative assessments

$\beta\gamma$ means the sum of beta emitters, gamma emitters, and low-toxicity alpha emitters
" means the sum of all other alpha emitters (i.e., other than low-toxicity alpha emitters)

The Basic Contamination Limits for All Packages: 49 CFR 173.443(a), Table 11	General Requirement:	Non-fixed (removable) contamination must be kept as low as reasonably achievable (ALARA)
	$\beta\gamma$: 0.4 Bq/cm² = 40 Bq/100 cm² = 1x10⁻⁵ µCi/cm² = 2200 dpm/100 cm²	
	": 0.04 Bq/cm² = 4 Bq/100 cm² = 1x10⁻⁶ µCi/cm² = 220 dpm/100 cm²	

The following exceptions and deviations from the above basic limits exist:

Deviation from Basic Limits	Regulation 49 CFR §§	Applicable Location and Conditions Which must Be Met:
10 times the basic limits	173.443(b) and 173.443(c) Also see 177.843 (highway)	On any external surface of a package in an **exclusive use shipment, during transport** including end of transport. Conditions include: (1) Contamination levels at beginning of transport must be below the basic limits. (2) Vehicle must not be returned to service until radiation level is shown to be ≤ 0.005 Mev/hr (0.5 mem./hr) at any accessible surface, and there is no significant removable (non-fixed) contamination.
10 times the basic limits	173.443(d) Also see 177.843 (highway)	On any external surface of a package, at the beginning or end of transport, if a closed transport vehicle is used, solely for transporting radioactive materials packages. Conditions include: (1) A survey of the interior surfaces of the empty vehicle must show that the radiation level at any point does not exceed 0.1 Mev/hr (10 mem./hr) at the surface, or 0.02 Mev/hr (2 mem./hr) at 1 meter (3.3 ft). (2) Exterior of vehicle must be conspicuously stenciled, **"For Radioactive Materials Use Only"** in letters at least 76 mm (3 inches) high, on both sides. (3) Vehicle must be kept closed except when loading and unloading.
100 times the basic limits	173.428	**Internal** contamination limit for **excepted package-empty packaging,** Class 7 (Radioactive) Material, shipped in accordance with 49 CFR 173.428. Conditions include: (1) The basic contamination limits (above) apply to **external** surfaces of package. (2) Radiation level must be ≤ 0.005 Mev/hr (0.5 mem./hr) at any external surface. (3) Notice in §173.422(a)(4) must accompany shipment. (4) Package is in unimpaired condition & securely closed to prevent leakage. (5) Labels are removed, obliterated, or covered, and the "empty" label (§172.450) is affixed to the package.

In addition, **after any incident** involving spillage, breakage, or suspected contamination, the modal-specific DOT regulations (§177.861(a), highway; §174.750(a), railway; and §175.700(b), air) specify that vehicles, buildings, areas, or equipment have "no significant removable surface contamination," before being returned to service or routinely occupied. The carrier must also notify offeror at the earliest practicable moment after incident.

Sample 2

HAZARDOUS MATERIAL SHIPPING CERTIFICATION

FOR COMPANY VEHICLE TRANSPORTING IRIDIUM 192 SEALED SOURCES

SHIPPER*	CONSIGNEE*
EZ Logging, Inc. 1234 Main Street Anywhere, USA 20000	Mo-Rad, Inc. 1234 Main Street Anywhere, USA 20000

DATE*		S.I. UNITS (CURIES)	TRANSPORT* INDEX (MR/HR @ 39 37")	CERTIFYING* SIGNATURE
35915	EZ Logging Pipe Services 4321 Broad Street Somewhere, USA			

DESCRIPTION OF PIECES AND CONTENTS

RQ RADIOACTIVE MATERIAL - SPECIAL FORM N.O.S. - UN 2974 - CLASS 7
YELLOW LABEL II - TRANSPORT INDEX NOT TO EXCEED 1.0
YELLOW LABEL II - TRANSPORT INDEX NOT TO EXCEED 10.0

LIQUID/GAS/SOLID (RQ RADIOACTIVE MATERIAL - N.O.S.7)	SPECIAL FORM	RADIOACTIVE MATERIAL EXCEPTED PACKAGE - LIMITED QUANTITY
☐ UN 2982	☐ UN 2974	☐ UN 2910

This is to certify that the above named materials are properly classified, described, packaged, marked, labeled and are in proper condition for transportation according to the applicable regulations of the DEPARTMENT OF TRANSPORTATION. (See certifying signature above)

INSTRUCTIONS

"Radioactive Yellow II Label" - 0.5 to 50 mR/hr on the surface of package and not over 1.0 mR/hr at 39.37" from container. Yellow II label does not require vehicle placards. NOTE: Do not transport if surface of container is over 50 mR/hr and/or over 1 mR/hr at 39.37" from container.

Shipping papers must be within reach of the driver when wearing a seat belt. Should the driver leave the vehicle, the shipping papers are to be left on the front seat of the driver's side or in a box on the driver's side of the vehicle.

If a motor vehicle accident occurs, it is required that an accident report be filed with the DOT within 15 days. Give no information regarding radioactive material to anyone present at the scene except police or DOT or NRC officials. Other information is to be obtained from the Radiation Safety Officer

---

EMERGENCY TELEPHONE NUMBER - 1-800-000-0000

---

* Substitute appropriate information for your device and shipment.

Appendix T

Model Waste Management Procedures

Model Waste Management Procedures

Model Waste Disposal Program

General Guidelines

A. All radioactivity labels must be defaced or removed from containers and packages prior to disposal. If nonradioactive waste is compacted, all radioactivity labels that are visible in the compacted mass must be defaced or removed.

B. Remind workers that non-radioactive waste should not be mixed with radioactive waste.

C. Occasionally monitor all procedures to ensure that radioactive waste is not created unnecessarily. Review all new procedures to ensure that waste is handled in a manner consistent with established operating and emergency procedures.

D. Evaluate the entire impact of various available disposal routes. Consider occupational and public exposure to radiation, other hazards associated with the material and routes of disposal (e.g., toxicity, carcinogenicity, pathogenicity, flammability), and costs.

E. Waste management program should include waste handling procedures. Also, procedures should be available and for well logging personnel who may collect waste from areas of use to bring to the storage area for eventual disposal.

Model Procedure for Disposal by Decay-in-Storage (DIS)

A. Only short-lived waste (physical half-life of less than or equal to 120 days) may be disposed of by DIS.

B. Short-lived waste should be segregated from long-lived waste (half-life greater than 120 days) at the source.

C. Waste should be stored in suitable well-marked containers, and the containers should provide adequate shielding.

D. Liquid and solid wastes must be stored separately.

E. When the waste container is full, it should be sealed. The sealed container should be identified with a label affixed or attached to it.

F. The identification label should include the date when the container was sealed, the longest-lived radioisotope in the container, date when ten half-lives of the longest-lived radioisotope will have transpired, and the initials of the individual who sealed the container. The container may be transferred to the DIS area.

G. The contents of the container should be allowed to decay for at least 10 half-lives of the longest-lived radioisotope in the container. The decay interval beginning at the time the radioactive waste container is sealed and placed in storage for DIS should be used for calculations and projected removal times.

H. Prior to disposal as ordinary trash, each container should be monitored as follows:

1. Check the radiation detection survey meter for proper operation.

2. Survey the contents of each container in a low background area.

3. Remove any shielding from around the container.

4. Monitor all surfaces of the container.

5. Discard the contents as ordinary trash only if the surveys of the contents indicate no residual radioactivity, i.e., surface readings are indistinguishable from background.

6. If the surveys indicate residual radioactivity, return the container to DIS area and contact the RSO for further instructions.

I. If the surveys indicate no residual radioactivity, record the date when the container was sealed, the disposal date, type of waste (used or unused material, gloves, etc.), survey instrument used, and the initials of the individual performing surveys and disposing of the waste.

Model Procedure for Disposal of Liquids into Sanitary Sewerage

A. Confirm that the liquid radioactive waste containing radioactive material being discharged is soluble or readily dispersible in water.

B. Calculate the amount of each radioisotope that can be discharged by using the information from prior, similar discharges and the information in 10 CFR 20, Appendix B.

C. Make sure that the amount of each radioisotope does not exceed the monthly and annual discharge limits specified in 10 CFR 20.2003(a)(4) and 10 CFR 20, Appendix B.

D. Record the date, radioisotope(s), estimated activity of each radioisotope, location where the material is discharged, and the initials of the individual discharging the radioactive waste.

E. Liquid radioactive waste must be discharged only via designated locations.

F. Discharge radioactive liquid waste slowly with water running from the faucet to dilute it.

G. Survey the designated disposal locations and surrounding work surfaces to confirm that no residual material or contamination remains.

H. Prior to leaving the area, decontaminate all areas or surfaces, if found to be contaminated.

I. Maintain disposal records that identify each radioisotope and its quantity and the concentration that is released into the sanitary sewer system.

Appendix U

Well Owner/Operator Agreement

Well Owner/Operator Agreement

TERMS AND CONDITIONS

For good and valuable consideration received, Customer (as identified on the face of this document) and [Insert Company Name] (hereafter "Insert Company Name Abbreviation") agree as follows:

A. CUSTOMER REPRESENTATION - Customer warrants that the well is in proper condition to receive the services, equipment, products, and materials to be supplied by [Insert Company Name Abbreviation]

B. PRICE AND PAYMENT - The services, equipment, products, and/or materials to be supplied hereunder are priced in accordance with [Insert Company Name Abbreviation] current price list. All prices are exclusive of taxes. If Customer does not have an approved open account with [Insert Company Name Abbreviation], all sums due are payable in cash at the time of performance of services or delivery of equipment, products, or materials. If Customer has an approved open account, invoices are payable on the [insert No.] day after the date of the invoice. Customer agrees to pay interest on any unpaid balance for the date payable until paid at the highest lawful contract rate applicable, but never to exceed [Insert No.]% per annum. In the event [Insert Company Name Abbreviation] employs an attorney for collection of any account, Customer agrees to pay attorney fees of [Insert No.]% of the unpaid account, plus all collection and court costs.

C. RELEASE AND INDEMNITY - CUSTOMER AGREES TO RELEASE [Insert Company Name Abbreviation] FROM ANY AND ALL LIABILITY FOR ANY AND ALL DAMAGES WHATSOEVER TO PROPERTY OF ANY KIND OWNED BY, IN THE POSSESSION OF, OR LEASED BY CUSTOMER AND THOSE PERSONS AND ENTITIES. CUSTOMER HAS THE ABILITY TO BIND BY CONTRACT. CUSTOMER ALSO AGREES TO DEFEND, INDEMNIFY AND HOLD [Insert Company Name Abbreviation] HARMLESS FROM AND AGAINST ANY AND ALL LIABILITY, CLAIMS, COSTS, EXPENSES, ATTORNEY FEES AND DAMAGES WHATSOEVER FOR PERSONAL INJURY, ILLNESS, DEATH, PROPERTY DAMAGE AND LOSS RESULTING FROM:

> LOSS OF WELL CONTROL; SERVICES TO CONTROL A WILD WELL WHETHER UNDERGROUND OR ABOVE THE SURFACE; RESERVOIR OR UNDERGROUND DAMAGE; DAMAGE TO OR LOSS OF OIL, GAS, OTHER MINERAL SUBSTANCES OR WATER; SURFACE DAMAGE ARISING FROM UNDERGROUND DAMAGE; DAMAGE TO OR LOSS OF THE WELL BORE; SUBSURFACE TRESPASS OR ANY ACTION IN THE NATURE THEREOF; FIRE; EXPLOSION; SUBSURFACE PRESSURE; RADIOACTIVITY; AND POLLUTION AND ITS CLEANUP AND CONTROL.

CUSTOMER'S RELEASE, DEFENSE, INDEMNITY AND HOLD HARMLESS OBLIGATIONS WILL APPLY EVEN IF THE LIABILITY AND CLAIMS ARE CAUSED BY THE SOLE, CONCURRENT, ACTIVE OR PASSIVE NEGLIGENCE, FAULT, OR STRICT LIABILITY OF ONE OR MORE MEMBERS OF THE [Insert Company Name Abbreviation], THE UNSEAWORTHINESS OF ANY VESSEL OR ANY DEFECT IN THE DATA PRODUCTS, SUPPLIES, MATERIALS OR EQUIPMENT FURNISHED BY [Insert Company Name Abbreviation]. [Insert Company Name Abbreviation] IS DEFINED AS [Insert Company Name Abbreviation] ITS PARENT, SUBSIDIARY, AND AFFILIATED COMPANIES AND ITS/THEIR OFFICERS, DIRECTORS, EMPLOYEES, AND AGENTS. CUSTOMER'S RELEASE, DEFENSE, INDEMNITY AND HOLD HARMLESS OBLIGATIONS APPLY WHETHER THE PERSONAL INJURY, ILLNESS, DEATH, PROPERTY DAMAGE OR LOSS IS SUFFERED BY ONE OR MORE MEMBERS OF THE [Insert Company Name Abbreviation], CUSTOMER, OR ANY OTHER PERSON OR ENTITY, AND THE CUSTOMER WILL SUPPORT SUCH OBLIGATIONS ASSUMED HEREIN WITH LIABILITY INSURANCE TO THE MAXIMUM EXTENT ALLOWED BY APPLICABLE LAW.

D. EQUIPMENT LIABILITY - Customer shall at its risk and expense attempt to recover any [Insert Company Name Abbreviation] equipment lost or lodged in the well. If the applicant is recovered and reputable, Customer shall pay the repair costs, unless caused by [Insert Company Name Abbreviation] sole negligence. If a radioactive source becomes lost or lodged in the well, Customer shall meet all requirements of Section 39.15(a) of the Nuclear Regulatory Commission regulations and any other applicable laws or regulations concerning retrieval or abandonment and shall permit [Insert Company Name Abbreviation] to monitor the recovery or abandonment efforts all at no risk or liability to [Insert Company Name Abbreviation]. Customer shall be responsible for damages to or loss of [Insert Company Name Abbreviation] equipment, products, and materials while in transit aboard Customer-applied transportation, even if such is arranged by [Insert Company Name Abbreviation] at Customer's request, and during loading and unloading from such transport. Customer will also pay for the repair or replacement of [Insert Company Name Abbreviation] equipment damaged by corrosion or abrasion due to well effluents.

E. LIMITED WARRANTY - [Insert Company Name Abbreviation] warranty only applies to the equipment, products, and materials supplied under this agreement and that same are free from defects in workmanship and materials for one year from date of delivery. THERE ARE NO WARRANTIES, EXPRESS OR IMPLIED, OF MERCHANTABILITY, FITNESS OR OTHERWISE BEYOND THOSE STATED IN THE IMMEDIATELY PRECEDING SENTENCE. [Insert Company Name Abbreviation] sole liability and Customer's exclusive remedy in any cause of action (whether in contract, tort, breach of warranty or otherwise) arising out of the sale, lease or use of any equipment, products, or materials is expressly limited to the replacement of such on their return to [Insert Company Name Abbreviation] or, at [Insert Company Name Abbreviation] option, to the allowance to Customer of credit for the cost of such items. In no event shall [Insert Company Name Abbreviation] be liable for special, incidental, indirect, consequential, or punitive damages. Because of the uncertainty of variable well conditions and the necessity of relying on fads and supporting services furnished by other, [Insert Company Name Abbreviation] IS UNABLE TO GUARANTEE THE EFFECTIVENESS OF THE EQUIPMENT, MATERIALS, OR SERVICE, NOR THE ACCURACY OF ANY CHART INTERPRETATION, RESEARCH ANALYSIS, JOB RECOMMENDATION OR OTHER DATA FURNISHED BY [Insert Company Name Abbreviation]. [Insert Company Name Abbreviation] personnel will use their best efforts in gathering such information and their best judgment in interpreting it, but Customer agrees that [Insert Company Name Abbreviation] shall not be liable for and CUSTOMER SHALL INDEMNIFY [Insert Company Name Abbreviation] AGAINST ANY DAMAGES ARISING FROM THE USE OF SUCH INFORMATION, even if such is contributed to by [Insert Company Name Abbreviation] negligence or fault. [Insert Company Name Abbreviation] also does not warrant the accuracy of data transmitted by electronic process, and [Insert Company Name Abbreviation] will not be responsible for accidental interception of such data by third parties.

F. GOVERNING LAW - The validity, interpretation and construction of this agreement shall be determined by the laws of the jurisdiction where the services are performed or the equipment or materials are delivered.

G. WAIVER - Customer agrees to waive the provisions of the Texas Deceptive Trade Practices-Consumer Protection Act or any similar Federal or State act to the extent permitted by law.

H. MODIFICATIONS - Customer agrees that [Insert Company Name Abbreviation] shall not be bound by any modifications to this agreement, except where such modification is made in writing by a duly authorized executive officer of [Insert Company Name Abbreviation]. Requests for modifications should be directed to [Insert Name and Title].

Appendix V

Actions to be Taken if a Sealed Source is Ruptured

Actions to be Taken if a Sealed Source is Ruptured

Paragraph 39.69(a) requires immediate initiation of emergency procedures if there is evidence that a sealed source has ruptured or that licensed materials have caused contamination.

Your procedures should instruct logging personnel to:

- Notify immediately the RSO or other appropriate management personnel.

- Notify the well owner or operator as soon as possible.

- Notify the NRC operations center at the telephone number specified in 10 CFR 20.2202(d)(2) — (301) 816-5100.

- Secure and restrict access to the area until responsible individuals arrive.

- Instruct individuals on site not to take any unnecessary actions that could spread contamination.

- Minimize inhalation or ingestion of licensed material by using protective clothing and respirators.

- Discuss procedures for preventing the spread of contamination and for minimizing inhalation or ingestion with any potentially exposed personnel.

- Obtain suitable radiation survey instruments as required by Section 39.33(b).

NRC FORM 335
(2-89)
NRCM 1102,
3201, 3202

U.S. NUCLEAR REGULATORY COMMISSION

BIBLIOGRAPHIC DATA SHEET

(See instructions on the reverse)

1. REPORT NUMBER
(Assigned by NRC, Add Vol., Supp., Rev., and Addendum Numbers, if any.)

NUREG 1556, Volume 14

2. TITLE AND SUBTITLE

Consolidated Guidance About Materials License: Program Specific Guidance about Well Logging, Tracer, and Field Flood Study Licensees

Final Report

3. DATE REPORT PUBLISHED

MONTH	YEAR
June	2000

4. FIN OR GRANT NUMBER

5. AUTHOR(S)

Jack E. Whitten, Steven R. Courtemanche, Andrea R. Jones
Richard E. Penrod, David B. Fogle

6. TYPE OF REPORT

Final

7. PERIOD COVERED (Inclusive Dates)

8. PERFORMING ORGANIZATION - NAME AND ADDRESS (If NRC, provide Division, Office or Region, U.S. Nuclear Regulatory Commission, and mailing address; if contractor, provide name and mailing address.)

Office of Nuclear Material Safety and Safeguards
Division of Industrial and Medical Nuclear Safety
U.S. Nuclear Regulatory Commission
Washington, DC 20555-0001

9. SPONSORING ORGANIZATION - NAME AND ADDRESS (If NRC, type "Same as above"; if contractor, provide NRC Division, Office or Region, U.S. Nuclear Regulatory Commission, and mailing address.)

SAME

10. SUPPLEMENTARY NOTES

11. ABSTRACT (200 words or less)

As part of its redesign of the materials licensing process, NRC is consolidating and updating numerous guidance documents into a single comprehensive repository as described in NUREG-1539, "Methodology and Findings of the NRC's Materials Licensing Process Redesign," dated April 1996, and draft NUREG-1541, "Process and Design for Consolidating and Updating Materials Licensing Guidance," dated April 1996. NUREG-1556, Vol. 14, "Consolidated Guidance about Materials Licenses: Program-Specific Guidance about Well Logging, Tracer, and Field Flood Study Licenses," dated May 2000, is the fourteenth program-specific guidance document developed for the new process and is intended for use by applicants, licensees, and NRC staff, and will also be available to Agreement States. This document combines and updates the guidance found in Draft Regulatory Guide, "Guide for the Preparation of Applications for the Use of Radioactive Materials in Well Logging Operations," dated July 1987. This report takes a more risk-informed, performance-based approach to licensing of well logging, tracer, and field flood study operations, and reduces the information (amount and level of detail) needed to support an application for these activities.

12. KEY WORDS/DESCRIPTORS (List words or phrases that will assist researchers in locating the report.)

Well Logging

13. AVAILABILITY STATEMENT

unlimited

14. SECURITY CLASSIFICATION

(This Page)
unclassified

(This Report)
unclassified

15. NUMBER OF PAGES

16. PRICE

www.ingramcontent.com/pod-product-compliance
Lightning Source LLC
Chambersburg PA
CBHW080239180526
45167CB00006B/2337